Casing and Liners for Drilling and Completion

Casing and Liners for Drilling and Completion

Editor

Aman Gupta

scitus
academics

Casing and Liners for Drilling and Completion

Edited by **Aman Gupta**

Printed in 2017

ISBN: 978-1-68117-339-9

Library of Congress Control Number: 2015939251

Notice

Contents

Preface

Casing and liners for Drilling and Completions, provides the engineer and well designer with up-to-date information on critical properties, mechanics, design basics and newest applications for today's type of well. Renovated and simplified to cover operational considerations, pressure loads, and selection steps, and gives you the knowledge to execute the essential and fundamental features of casing and liners.

Editor

Copolymer SJ-1 as a Fluid Loss Additive for Drilling Fluid with High Content of Salt and Calcium

Hongping Quan[1,2], Huan Li[1,2], Zhiyu Huang[1,2], Tailiang Zhang[1,2], and Shanshan Dai[1,2]

[1]School of Chemistry and Chemical Engineering, Southwest Petroleum University, Chengdu 610500, China

[2]Department of Education, Engineering Research Center of Oilfield Chemistry, Chengdu 610500, China

ABSTRACT

A ternary copolymer of 2-acrylamide-2-methyl propane sulfonic acid (AMPS), acrylamide (AM), and allyl alcohol polyoxyethylene ether (APEG) with a side chain polyoxyethylene ether $(C_2H_4O)_n$ SJ-1 were designed and synthesized in this work. Good temperature resistance and salt tolerance of "$-SO_3^-$" of AMPS, strong absorption ability of "amino-group" of AM, and good hydrability of side chain polyoxyethylene ether $(C_2H_4O)_n$ of APEG provide SJ-1 excellent properties as a fluid loss additive. The chemical structure of ternary copolymer was characterized by Fourier transform infrared (FTIR) spectroscopy. The molecular weight and its distribution were determined by gel permeation chromatography (GPC). The API fluid loss of drilling fluid decreased gradually with the increasing concentration of NaCl and $CaCl_2$ in the mud system. SJ-1 was applied well in the drilling fluid even at a high temperature of 220°C. Results of zeta potential of modified drilling fluid showed the dispersion stability of drilling fluid system. Scanning electron microscopy (SEM) analysis showed the microstructure of the surface of the filter cake obtained from the drilling fluid modified by SJ-1.

INTRODUCTION

Water-based drilling muds including bentonite were well-known and widely used in the petroleum industry recently [1–4]. The main components of water-based drilling fluid are water, salt, insert solids, and clays such as montmorillonite. Among the important functions of water-based drilling fluid were to form filter cake on the wall of the well bore, prevent water leakage, and maintain the stability of the well wall [5]. The properties of the water-based drilling fluid, such as the rheology and filtration loss, are affected by the fluid loss additive. Polymers, which are nontoxic, degradable, and environment friendly [6–8], are the best choice to be used as drilling fluids additives. However, the traditional polymers as fluid loss additive are not stable at the work condition of drilling fluid, especially at high salt condition and high temperatures. The resulting drilling fluids often present poor properties such as high filtration. If the drilling fluid was applied in the high salt or calcium environment, the flocculated structure of clay particles will appear and the value of API fluid loss will increase obviously. Furthermore downhole accident such as borehole collapse will happen in severe case. To solve the problems mentioned above, a novel fluid loss additive SJ-1 was developed in this work [2, 9–11].

To make sure the fluid loss additive has an excellent ability of absorption, the monomer with strong absorption groups, such as amino-group, are introduced [12, 13]. To make sure the additive has an excellent ability of salt tolerance, the monomers with stable group which is not sensitive to the cation, such as $-SO_3^-$, are introduced [14–16]. To make the fluid loss additive have an excellent ability of temperature resistance, the monomers with inflexibility groups such as big side group are introduced. At the same time, the monomers that are hard to hydrolyze are introduced to get rid of the side effect hydrolysis, such as polyoxyethylene ether $(C_2H_4O)_n$ [17]. In this work, a good property of filtrate loss control even at high temperatures and salt condition of SJ-1 is synthesized by polymerization of AMPS, AM, and APEG.

EXPERIMENTAL

Materials

AMPS and AM were from Chengdu Kelong Chemical Reagent Factory (China). APEG-1000 were of industrial grade from Jiangsu Haian petrochemical factory (China). Potassium persulfate, sodium hydroxide, and sodium bisulfite were analytical grade from Chengdu Kelong Chemical Reagent Factory (China).

Synthesis

APEG (0.006 mol) was added in a three-necked flask, using moderate amount of water to dissolve in a heated water bath. And then AMPS (0.029 mol) and AM (0.084 mol) were added to the mixed solution. The pH was adjusted to 7-8 with NaOH. Finally, 0.2 wt% potassium persulfate and sodium bisulfite were added into the system. The reaction was performed under 60°C with stir speed of 300 rmp. After completion of the reaction, the product was extracted with ethanol, shear granulation, vacuum drying, and grinding to obtain a white powdery polymer fluid loss additive. Chemical structure of SJ-1 was shown in Scheme 1.

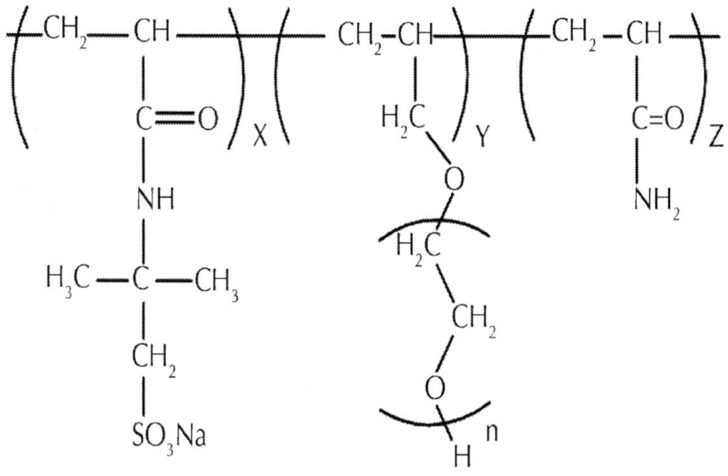

Scheme 1: Chemical structure of SJ-1.

In laboratory studies, the best parameters of copolymer SJ-1 were obtained: monomer concentration was 15 wt%; molar ratio of n(AM) : n(AMPS) : n(APEG) was 14 : 5 : 1; the reaction temperature was 60°C; the concentration of monomer initiator was 0.2 wt%.

Characterization

The pellet samples were prepared by pressing the mixture of the SJ-1 and KBr and then measured with FTIR (WQF-520, China) spectrophotometer at range between 4000 and 500 cm^{-1}.

The molecular weight distribution of SJ-1 was measured by GPC (Waters e2695, USA). The polymer was dissolved into distilled water forming the solution with concentration of 2 mg/mL. The measurement was performed at the room temperature (23°C) for 90 min.

Sample Preparation

Freshwater base mud containing 4 wt% of sodium bentonite and 0.2 wt% of Na_2CO_3 was prepared by mixing the raw bentonite,

Na_2CO_3, and freshwater at a certain ratio, stirring for 20 min at a high speed of 10000 rpm, and aging for 24 h at room temperature. Salt-water base mud was prepared by adding different concentration of NaCl into the above freshwater base mud and then submitted to a prehydration period of 24 h. Calcium-water base mud was prepared by adding different concentration of $CaCl_2$ into the above freshwater base mud and then submitted to a prehydration period of 24 h. Polymer based mud was prepared by adding different concentration of SJ-1 into the freshwater base mud, salt-water base mud, and calcium-water base mud, respectively, and then submitted to a prehydration period of 24 h.

Fluid Filtration Property Test

Drilling fluid filtrations were measured according to American Petroleum Institute (API) specifications and Chinese SY/T5621-93. The API filtrate volume (FLAPI) of the mud was determined with a medium-pressure filtration apparatus (ZNS-2 type, China).

Apparent viscosity (AV), plastic viscosity (PV), and yield point (YP) were measured by rotational viscometer (ZNN-D6, China) at different temperatures.

Aging experiments of bentonite-polymer fluids were carried out in a frequency conversion rolling oven (BRGL-7 type, China) at a series of temperatures for 16 h.

The filter cake of the sample was tested by SEM (JSM-7500F, Japan) analysis and the filtrate of the sample was tested by zeta electric potential (Zeta PALS/90plus, Brookhaven, America) at different conditions.

RESULTS AND DISCUSSION

Chemical Structure and Molecular Weight

The chemical structure of SJ-1 was analyzed by FTIR after being purified; the results were shown in Figure 1.

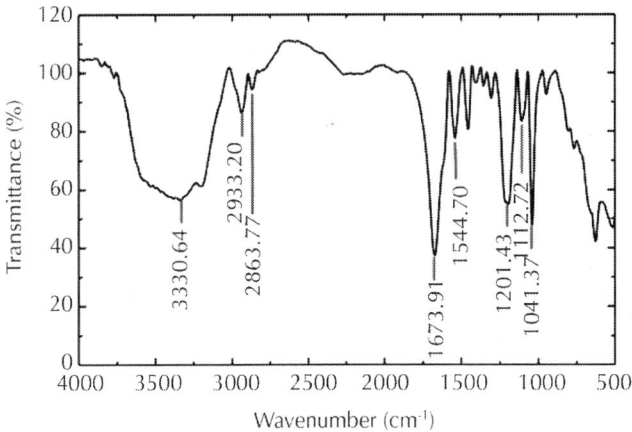

Figure 1: FTIR spectrum of polymerization product.

A strong absorption peak of 3330.64 cm⁻¹ was assigned to the stretching vibration of N–H. The absorption peaks observed at 2933.20 cm⁻¹ and 2863.77 cm⁻¹ were due to the stretching vibration of –CH₃ and –CH₂, respectively. The absorption peak of 1673.91 cm⁻¹ was attributed to the stretching peak of C=O. The absorption peak of 1544.70 cm⁻¹ was due to the stretching vibration of C–N, while 1201.43 cm⁻¹ was attributed to C–O–C. Moreover, absorption peaks of 1112.72 cm⁻¹ and 1041.37 cm⁻¹ were according to the bending vibration of –SO₃⁻. Results mentioned above were consistent with the chemical structure of SJ-1 shown in Scheme 1.

The molecular weight and its distribution of SJ-1 which was synthesized under the best optimum synthesis condition mentioned in Section 2.2 were determined by GPC. The result showed that the weight average molecular weight was 8.1×10^5, and the number average molecular was 6.2×10^5. Thus, the molecular weight distribution coefficient is 1.29. These all meet the technical requirements for oilfield application [18].

Effect of Concentration of SJ-1 on the Properties of Drilling Fluid

Different concentration of SJ-1 was added to the freshwater base mud. The rheological and API fluid loss properties have been measured by rotational viscometer and medium-pressure filtration apparatus, respectively. The results were shown in Table 1.

Table 1: Relationship between SJ-1 concentration and fluid loss

SJ-1 concentration/ wt%	AV/mPa·s	PV/ mPa·s	YP/Pa	API filter loss/mL
0.0	12.5	7.0	5.62	23.0
0.1	16.0	10.0	6.13	11.8
0.3	19.5	13.0	6.64	11.2
0.6	23.5	16.0	7.67	10.4
0.9	30.0	21.0	9.20	9.4
1.2	37.0	26.0	11.24	9.2
1.5	44.0	31.0	13.29	9.2
1.8	52.5	36.0	16.86	9.1

According to Table 1, with the increase of loading of SJ-1, the API fluid loss of drilling fluid decreased obviously. The values of fluid loss became stable, when the concentrations of SJ-1 was no less than 1.2 wt%. At the same time, the rheological property of drilling fluid changed as the increase of polymer concentration. SJ-1 contributes to build up the network structure, which results in a viscosity-building ability. If the rheological property of drilling fluid was bad, thinner might be required for the drilling fluid, so the optimum concentration of fluid loss addictive SJ-1 in the freshwater base mud was 1.2 wt%.

Evaluation of Temperature Resistance

1.2 wt% SJ-1 was added to the freshwater base mud, then, rolling in frequency conversion rolling oven for 16 h at different levels of temperature. The rheological and API fluid loss have been measured by rotational viscometer and medium-pressure filtration apparatus, respectively. The results were shown in Table 2.

Table 2: Fluid loss of drilling fluid changed with aging temperature

Experimental conditions	AV/mPa·s	PV/mPa·s	YP/Pa	API filter loss/ mL
120°C, 16 h	43.5	34.5	9.20	10.6
140°C, 16 h	38.5	32.0	6.64	10.6
160°C, 16 h	20.8	17.0	3.83	11.5
180°C, 16 h	13.0	9.5	3.58	12.1
200°C, 16 h	13.0	11.0	2.04	12.5
220°C, 16 h	12.5	11.0	1.53	13.0

The value of API fluid loss of drilling fluid raised with the increase of temperature, when the concentration of SJ-1 was equal to 1.2 wt%. The API fluid loss of drilling fluid was still below 13 mL even at 220°C showing its property of temperature resistant. Three reasons account for this phenomenon. Firstly, the main chain of the polymeric molecule is connected by C–C, which is stable at high temperatures. Secondly, the $-SO_3^-$ of the AMPS has a strong temperature resistance performance, introducing the $-SO_3^-$ into the polymer molecule can significantly improve the property of temperature resistance. Lastly, APEG molecule contains a polyoxyethylene group, which can improve the hydrophilic property and the adsorption capacity of the SJ-1 at high temperatures.

With the increase of temperature, the rheological property of drilling fluid which reflected on apparent viscosity, plastic viscosity, and yield value decreased. Some of the SJ-1 was degraded at high temperature, which can lead to breaking up the network structure to a certain extent. But this negative influence held within limits and control. The little change on the performance of the drilling fluid was less affected.

Evaluation of Salt Tolerance

The Effect of NaCl at 120°C

1.2 wt% SJ-1 and different concentration of NaCl were added to the freshwater base mud and, then, aged 16 h at 120°C. The rheological and API fluid loss have been measured by rotational viscometer and medium-pressure filtration apparatus, respectively. The results were shown in Table 3.

Table 3: Performance evaluation of salt resistance of drilling fluid at high temperature

Experimental conditions	NaCl concentration/ wt%	AV/ mPa·s	PV/ mPa·s	YP/Pa	**API filter loss/mL**
Aging temperature 120°C, aging time 16 h	0.0	32.0	27.5	4.60	10.6
	2.0	20.0	17.5	2.56	16.0
	5.0	17.5	15.0	2.56	13.0
	10.0	18.0	16.0	2.04	12.6
	15.0	17.0	15.0	2.04	12.6
	20.0	17.0	15.0	2.04	12.6
	25.0	17.5	15.0	2.56	10.4
	30.0	19.5	18.0	1.53	7.5

Table 3 illustrated that the API fluid loss of drilling fluid decreased gradually by the growing concentration of NaCl in the mud system. When the salt concentration was 30 wt%, the fluid loss achieved the lowest level, namely 7.5 mL, which indicated the good salt tolerance of SJ-1.

Five reasons account for this phenomenon. Firstly, when the NaCl was added to the freshwater base mud, the value of the zeta electric potential of the clay particles would decrease and the hydration shell would be reduced, which would lead to generating the flocculated structure in the freshwater base mud. So, the API filter loss increases. Secondly, when the NaCl was added to the freshwater base mud, the adsorption of the fluid loss additive on the surface of the clay particles would increase, which would lead to the increase of the zeta electric potential of the clay particles and the thickening of hydration shell. So, the API filter loss would decrease. When the concentration of the NaCl was 2 wt%, the effect of the NaCl on the clay particles played a leading role, and the API filter loss was increased. Meanwhile, with the increase of the concentration of the NaCl, the effect of the NaCl on the fluid loss additive was increased and the API filter loss was decreased. When the concentration of the NaCl was 25 wt%, the effect of the

NaCl on the fluid loss additive played a leading role, so the API filter loss was lower than that without the NaCl. Thirdly, the SJ-1 with good temperature resistant remains at a comparatively high level. And it is beneficial for adsorbing clay particles and free water to form aggregate, which forms a thin and dense mud cake on the wall. The appearance of mud cake effectively reduces the filtrate loss. Lastly, because of the adding of fluid loss additive SJ-1, the hydrophilic radical of this treating chemical mainly includes sulfonic acid and polyoxyethylene side chain which enjoy a favorable hydrophilicity and good solubility in a saline environment. It can bring enough hydrated film to clay and ensure the drilling fluid system has good salt tolerance.

The Effect of $CaCl_2$ at 120°C

1.2 wt% SJ-1 and different concentration of $CaCl_2$ were added to the freshwater base mud and, then, aged 16 h at 120°C. The rheological and filtration properties have been measured by rotational viscometer and medium-pressure filtration apparatus, respectively. The results have been shown in Table 4.

Table 4: Performance evaluation of calcium resistance of drilling fluid under high temperature

Experimental conditions	$CaCl_2$ concentration/ wt%	AV/ mPa·s	PV/ mPa·s	YP/Pa	API filter loss/mL
Aging temperature 120°C, aging time 16 h	0.0	32.0	27.5	4.60	10.6
	2.0	9.5	9.0	0.51	26.0
	4.0	10.0	9.0	1.02	19.0
	6.0	9.0	8.0	1.02	17.0
	8.0	8.5	8.0	0.51	12.6
	10.0	8.5	8.0	0.51	12.0

Table 4 illustrated that the fluid loss of drilling fluid decreased gradually by the growing number of $CaCl_2$ in the mud system. When the $CaCl_2$ was 8 wt%, the fluid loss was lower than 12.6 mL, however the API fluid loss remain unchanged with the increase of $CaCl_2$. When the $CaCl_2$ was 10%, the API fluid loss was 12.0 mL.

Two reasons account for this phenomenon. Firstly, the $-SO_3^-$ of the SJ-1 molecular chain has strong hydration capacity and the hydration of the clay particles can be thickened. Meanwhile, the $-SO_3^-$ and the calcium do not cause the precipitation reaction in the drilling fluid containing calcium ions. So the SJ-1 can form a thick hydration shell on the surface of clay, which can ensure that the SJ-1 has good fluid loss performance in the drilling fluid containing calcium ions. Secondly, because of the $-SO_3^-$, the SJ-1 can resist a temperature of 120°C in which polymer rarely degrades or degrades in a small quantity; the original nature of many polymer molecules can be maintained, and they can adsorb clay particles and free water, forming an aggregate that acts as a thin and dense mud cake on the borehole wall, which effectively reduces the leakage of filtrate into formation.

Zeta Electric Potential Test

1.2 wt% SJ-1 was added to the freshwater base mud in different temperatures. The zeta electric potential and API filter loss have been measured by zeta potential analyzer and medium-pressure filtration apparatus, respectively. The results were shown in Figure 2.

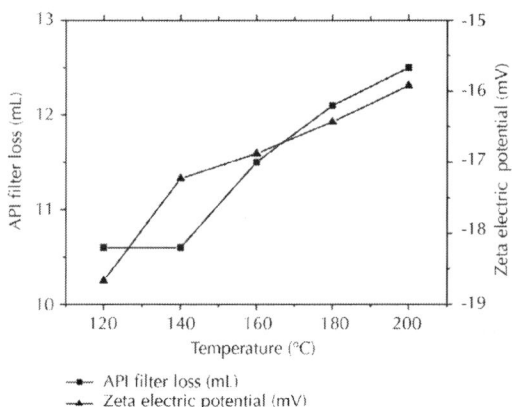

Figure 2: The impact of zeta electric potential and API filter loss on mud system in different temperatures.

Figure 2 shows that, in different temperatures, both drilling fluid system and zeta electric potential have a big change. The absolute value of the zeta electric potential of the mud system gradually decreased with the increase of aging temperature, and the filter loss gradually increases with the increase of experimental temperatures. It means that with the increase of aging temperatures, the character of mud system changes a lot, in which the absorption capacity of polymers on the surface of clay particles is reducing, leading to the increase of filter loss of mud. However, because the SJ-1 contains many $-SO_3^-$ and side chain polyoxyethylene ether $(C_2H_4O)_n$ which exhibits unique hydration and dispersion capacities at high temperature, the negative influence held within limits. The zeta electric potential test shows that this mud system still has a good dispersion stability in high temperature.

1.2 wt% SJ-1 and different concentration of NaCl were added to the freshwater base mud at 25°C. The zeta electric potential and API filter loss have been measured by zeta potential analyzer and medium-pressure filtration apparatus, respectively. The results were shown in Figure 3.

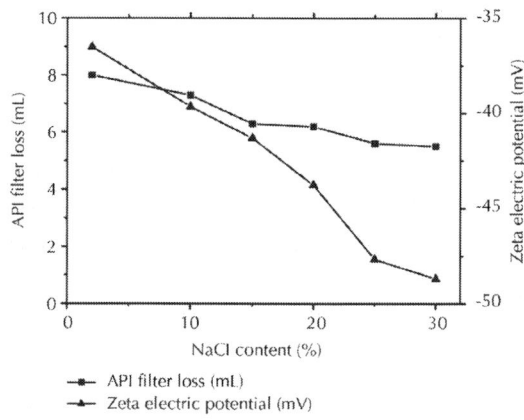

Figure 3: The impact of zeta electric potential and API filter loss on mud system in different salt contents.

Figure 3 shows that, with an increase of NaCl, filter loss is decreasing, along with the absolute value of zeta electric potential rising up, which is compatible with the data of zeta electric potential test. With the increase of electrolyte, the coalescence of electrolyte is gradually gentle and the aggregation of clay particles in drilling fluid is impeded, resulting in the increase of the small particles in the mud with the expansion of the degree of mineralization. At the same time, SJ-1 particles are adsorbed on the surface of clay particles; sulfonic acid group can chemically react to Na^+ and Ca^{2+} in salt; the steric effect of polyoxyethylene side chain in particles is large. Thus, it is more difficult to aggregate clay particles, and the compression of salt on the electric double layer is reduced, with the improvement of the resistance of fluid loss additive to salt. In addition, with the increase of polymer, both apparent and plastic viscosities are increasing to a small extent.

SEM Analysis of Filter Cake

In order to explore the microstructure of the filter cake formation and the filtration mechanism research, the formation of the filter cake was observed by type JSM7500F SEM.

Freshwater base mud and polymers of 1.2 wt% density are added into drilling fluid to test the API filter loss in normal temperature. After drying, the appearance of filter cake can be seen through SEM in Figure 4.

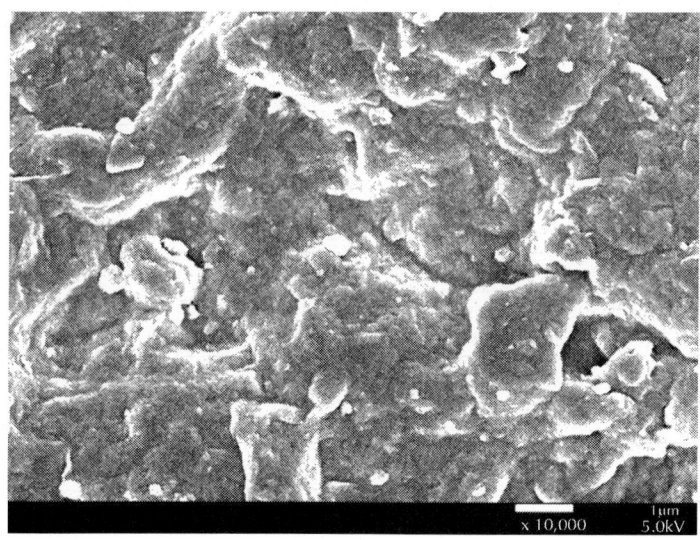

(a)

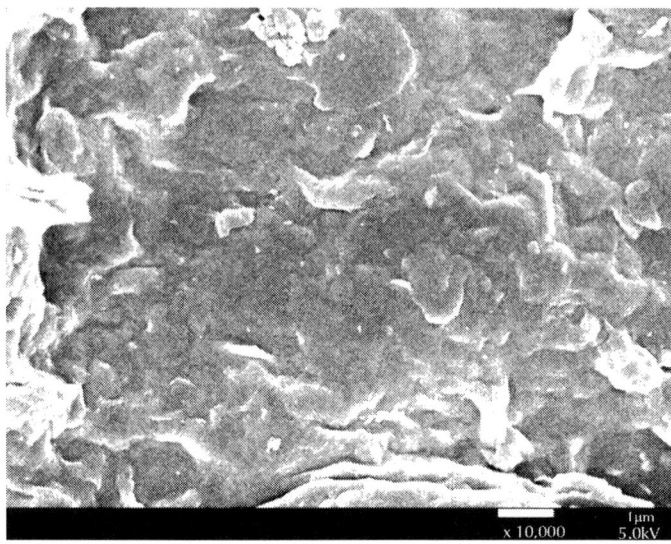

(b)

Figure 4: SEM photos of (a) API filter cake formed by base mud and (b) API filter cake modified with SJ-1 (both of aging temperature 120°C).

Figure 4(a) shows that the surface of the base mud is uneven and friable, with some ravines and small poles. Some big particles can be seen, so the mud is bad in dispersion and the cake formed is of a poor quality, causing a terrible dehydration.

Figure 4(b) shows that after an aging process, the cakes are uniform and dense on their surfaces, with no big pores and coarse particles. Polymer adsorbed on clay particles, forming a reticulated polymer, which effectively blocks the pore of borehole wall and prevents the filtrate in the drilling fluid leaking to formation. Thus, the drilling fluid is presented to be in good character and the filtrate loss is less than base mud. Then, hydration swelling and dispersion of shale are decreasing, which can better avoid borehole wall collapsing to protect reservoir.

Freshwater base mud polymers of 1.2% density are added into freshwater base mud, which is then stirred up speedily for 10 minutes. Next, NaCl is mixed into it and then stirred up for 10 minutes. API filtrate loss is tested and Figure 5 shows the SEM images of the filter cake.

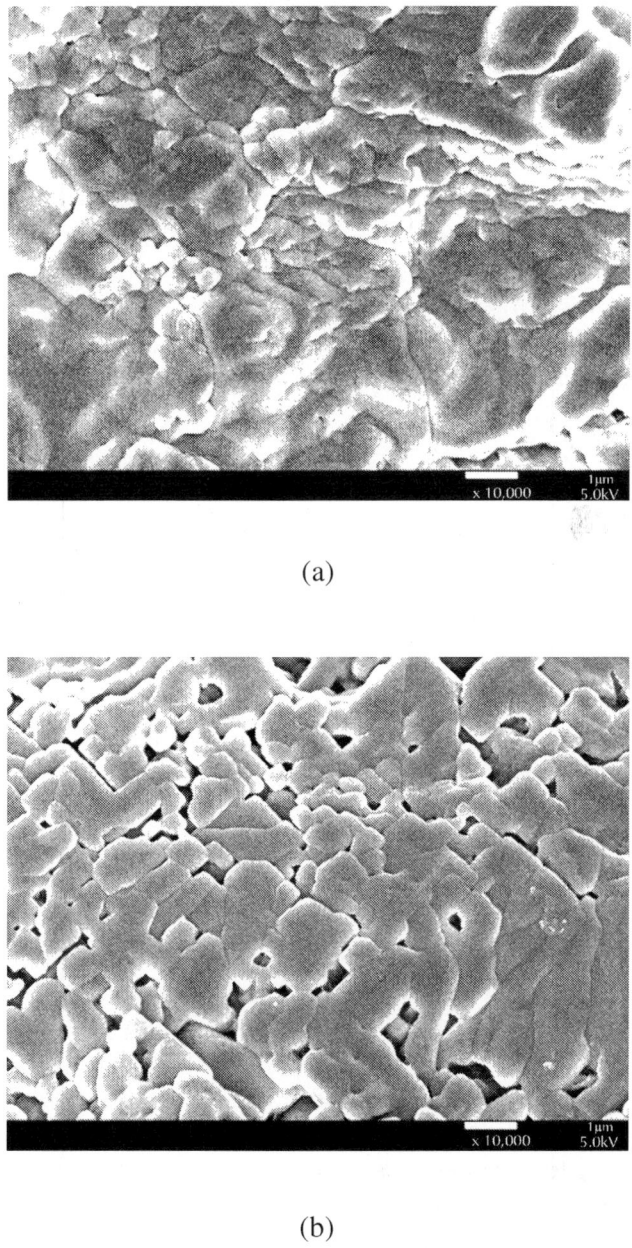

(a)

(b)

Figure 5: SEM photos of (a) API filter cake formed by base mud + SJ-1 + 30 wt% NaCl and (b) API filter cake formed by base mud + SJ-1 + 30 wt% NaCl after the aging process of 30 min.

Figure 5(a) shows that the filter cakes, with NaCl of 30% density added into the mud system mixed with the fluid loss additive, are dense on their surfaces and distributed regularly in tight connection, narrowing the pores for less filtration loss. It means that filtration loss of drilling fluid can be effectively reduced by adding a large quantity of NaCl into drilling fluid system. The polyoxyethylene side chain in the fluid loss additive raises the resistance of polymers to salt, and its plastic-protection function impede the coalescence of NaCl to clay particles, contributing to the regular distribution of clay particles in the drilling fluid system. Thus, the cake formed is dense.

Figure 5(b) shows that the cakes, with NaCl of 30% density added into the mud system mixed with the fluid loss additive aging 30 min at 120°C, are dense and smooth on their surfaces and are distributed regularly in tight connection, narrowing the pores of filtrate loss.

CONCLUSIONS

With FTIR analysis, the composition of the synthetic polymer SJ-1 was consistent to the designed structure. The weight average molecular weight of SJ-1 was 8.1×10^5, and its number average molecular was 6.2×10^5. The molecular weight distribution coefficient was 1.29. The character evaluation proves that the best amount of SJ-1 in the mud is 1.2%, and the API filtrate loss is 9.2 mL. Salt resistance of SJ-1 is effective and keeps its effectiveness in the drilling fluid system in which the aging temperature is 220°C, the mass density of NaCl is 30 wt%, and $CaCl_2$ is 10 wt%. The zeta electric potential test on clay particles proves that with an increase of SJ-1, the API filtrate loss in drilling fluid is decreasing and the absolute value of zeta electric potential is increasing. Hydrated membrane is thickened by the increase of the density of negative charge of clay particles, contributing to the reduction of filtrate loss of drilling fluid. In addition, the SEM analysis proves that cakes with SJ-1 added in are even and dense on their surfaces with no big pores or coarse particles, and the filtrate loss is less than base mud. Besides, those cakes can effectively resist the pollution of electrolytes. And in the mud system with NaCl of 30% density, the cakes formed are distributed regularly in tight connection, narrowing the filtrate loss pores and reducing the amount of filtrate loss.

ACKNOWLEDGMENT

The authors thank the Engineering Research Center of Oilfield Chemistry, Ministry of Educational Key for experiment support.

REFERENCES

1.   L. Ali and M. A. Barrufet, "Using centrifuge data to investigate the effects of polymer treatment on relative permeability," Journal of Petroleum Science and Engineering, vol. 29, no. 1, pp. 1–16, 2001.

2.   T. Wan, J. Yao, S. Zishun, W. Li, and W. Juan, "Solution and drilling fluid properties of water soluble AM-AA-SSS copolymers by inverse microemulsion," Journal of Petroleum Science and Engineering, vol. 78, no. 2, pp. 334–337, 2011.

3.   M. V. Kok, "Rheological and thermal analysis of bentonites for water base drilling fluids," Energy Sources A: Recovery, Utilization, and Environmental Effects, vol. 26, no. 2, pp. 145–151, 2004.

4.   M. V. Kok, "A rheological characterization and parametric analysis of a bentonite sample," Energy Sources A: Recovery, Utilization and Environmental Effects, vol. 33, no. 4, pp. 344–348, 2011.

5.   Y. Zhuang, Z. Zhu, H. Chao, and B. Yang, "Effect of added salt on properties of aqueous SMP solution,"Journal of Applied Polymer Science, vol. 55, no. 7, pp. 1063–1067, 1995.

6.   M. V. Kok, "Thermal characterization of sepiolite samples," Energy Sources A: Recovery, Utilization and Environmental Effects, vol. 35, no. 2, pp. 173–183, 2013.

7.   M. V. Kok, "Statistical approach of two-three parameters rheological models for polymer type drilling fluid analysis," Energy Sources A: Recovery, Utilization and Environmental Effects, vol. 32, no. 4, pp. 336–345, 2009.

8.   M. V. Kok, "Thermal analysis and rheological study of ocma type bentonite used in drilling fluids,"Energy Sources A: Recovery, Utilization and Environmental Effects, vol. 35, no. 2, pp. 122–133, 2013. ·

9.   C. Collette, F. Lafuma, R. Audebert, and R. Brouard, "Macromolecular systems in heat-resistant drilling fluids; advantages of gels on linear polymers," Journal of Applied Polymer Science, vol. 53, no. 6, pp. 755–762, 1994.

10. V. C. Kelessidis, C. Papanicolaou, and A. Foscolos, "Application of Greek lignite as an additive for controlling rheological and filtration properties of water-bentonite suspensions at high temperatures: a review," International Journal of Coal Geology, vol. 77, no. 3-4, pp. 394–400, 2009.

11. R. Caenn and G. V. Chillingar, "Drilling fluids: state of the art," Journal of Petroleum Science and Engineering, vol. 14, no. 3-4, pp. 221–230, 1996.

12. J. Plank, N. R. Lummer, and F. Dugonjić-Bilić, "Competitive adsorption between an AMPSV-based fluid loss polymer and Welan gum biopolymer in oil well cement," Journal of Applied Polymer Science, vol. 116, no. 5, pp. 2913–2919, 2010.

13. J. Y. Ma, H. L. Zheng, M. Z. Tan et al., "Synthesis, characterization, and flocculation performance of anionic polyacrylamide P (AM-AA-AMPS)," Journal of Applied Polymer Science, vol. 129, no. 4, pp. 1984–1991, 2013.

14. C. Tiemeyer and J. Plank, "Synthesis, characterization, and working mechanism of a synthetic high temperature (200°C) fluid loss polymer for oil well cementing containing allyloxy-2-hydroxy propane sulfonic (AHPS) acid monomer," Journal of Applied Polymer Science, vol. 128, no. 1, pp. 851–860, 2013. ·

15. Y. Liu, W. P. Gates, A. Bouazza, and R. K. Rowe, "Fluid loss as a quick method to evaluate hydraulic conductivity of geosynthetic clay liners under acidic conditions," Canadian Geotechnical Journal, vol. 51, no. 2, pp. 158–163, 2014.

16. Q. Xiao, W. F. Xiao, and X. X. Liu, "A novel cement fluid loss additive P1402," Advanced Materials Research, vol. 941–944, pp. 1203–1207, 2014.

17. S. Akimoto, S. Honda, and T. Yasukohchi, "Polyoxyalkylene alkenyl ether-maleic ester copolymer and use thereof," US Patent 5142036, 1989.

18. J. P. Plank, "Water-based muds using synthetic polymers developed for high temperature drilling," Oil and Gas Journal, vol. 90, no. 9, pp. 40–45, 1992.

Chapter 2

Microbial Diversity and Methanogenic Activity of Antrim Shale Formation Waters from Recently Fractured Wells

Cornelia Wuchter[1], Erin Banning[1], Tracy J. Mincer[1], Nicholas J. Drenzek[2], and Marco J. L. Coolen[1]

[1]Marine Chemistry and Geochemistry Department, Woods Hole Oceanographic Institution, Woods Hole, MA, USA

[2]Reservoir Geosciences Department, Schlumberger Doll Research, Cambridge, MA, USA

ABSTRACT

The Antrim Shale in the Michigan Basin is one of the most productive shale gas formations in the U.S., but optimal resource recovery strategies must rely on a thorough understanding of the complex biogeochemical, microbial, and physical interdependencies in this and similar systems. We used Illumina MiSeq 16S rDNA sequencing to analyze the diversity and relative abundance of prokaryotic communities present in Antrim shale formation water of three closely spaced recently fractured gas-producing wells. In addition, the well waters were incubated with a suite of fermentative and methanogenic substrates in an effort to stimulate microbial methane generation. The three wells exhibited substantial differences in their community structure that may arise from their different drilling and fracturing histories. Bacterial sequences greatly outnumbered those of archaea and shared highest similarity to previously described cultures of mesophiles and moderate halophiles within the Firmicutes, Bacteroidetes, and δ- and ε-Proteobacteria. The majority of archaeal sequences shared highest sequence similarity to uncultured euryarchaeotal environmental clones. Some sequences closely related to cultured methylotrophic and hydrogenotrophic methanogens were also present in the initial well water. Incubation with methanol and trimethylamine stimulated methylotrophic methanogens and resulted in the largest increase in methane production in the formation waters, while fermentation triggered by the addition of yeast extract and formate indirectly stimulated hydrogenotrophic methanogens. The addition of sterile powdered shale as a complex natural substrate stimulated the rate of methane production without affecting total methane yields. Depletion of methane indicative of anaerobic methane oxidation (AMO) was observed over the course of incubation with some substrates. This process could constitute a substantial loss of methane in the shale formation.

INTRODUCTION

Microbial gas formation through decomposition of sedimentary organic matter (OM) comprises roughly 20% of the world's natural gas resources (Rice, 1992), and it is estimated that even more microbial gas is retained in the corresponding source rocks of unconventional biogenic gas shales (Milkov, 2011). In the U.S. unconventional gas resources account for nearly 10% of the total natural gas generation (Martini et al., 2003). The Michigan Basin, centered on the lower peninsula of Michigan in the U.S., is home of the Devonian (~380 Ma-old) Antrim Shale formation, one of the most productive gas shale formations in the U.S. The finely laminated Antrim Shale contains thermally immature OM with a total organic carbon (TOC) content of 0.5–24% (Shurr and

Ridgley, 2002). In the central and eastern Michigan basin the Antrim shale contains thermogenic gas, created by pressure and thermal cracking of OM. Here, gas production rates in drilled wells are low and economically not successful (Martini et al., 2008). In contrast, shale gas production along the northern margin of the Michigan Basin is high and became a target area of rapid development starting with 100 wells in 1985 to over 12,000 gas-and water producing wells installed today (Martini et al., 2004; DEQ, 2012). Previous geochemical studies provided evidence for biologically mediated methane generation in the recent geological past along the northern margin of the Michigan Basin (Martini et al., 1996). Dilution of deep basin brines with meteoric waters and deep groundwater recharge during Pleistocene glaciations resulted in a steep salinity gradient in Antrim Shale pore waters with extremely diluted water along the margins to greater than 5 M NaCl at the center of the basin (McIntosh et al., 2002). The Antrim shale has been hydrologically isolated from surface water for at least 7000 years (Martini et al., 1998; McIntosh et al., 2002) and the formation water lacks abundant inorganic electron acceptors other than carbon dioxide, including sulfate and iron oxyhydroxides (Waldron et al., 2007). Such conditions are favorable for methanogens which often thrive in environments where carbon dioxide is the sole available electron acceptor (Whitman et al., 2006). Although the high salt concentrations in the Antrim formation water inhibit many microorganisms, halophilic organisms can thrive under a wide range of NaCl concentrations (Oren, 1999). A recent study on enrichment cultures from formation water of the methane-generating zone of the Antrim shale provided evidence of halophilic methanogenic communities growing at up to 2.5 M NaCl concentrations (Waldron et al., 2007), indicating that active methanogenesis is an ongoing process in the northern margins of the Antrim gas shale.

Although little is known of microbial communities in gas shale formations, similarly hydrocarbon-rich, anaerobic environments such as petroleum reservoirs and subsurface coal beds have been relatively well-studied (Magot et al., 2000; Strąpoć et al., 2008, 2011). Similar to the Antrim Shale, carbon dioxide is the predominantly bioavailable electron-acceptor in coal bed formation water (Strąpoć et al., 2011). Several studies have demonstrated the presence of active methanogenic archaea in coal beds (Shumkov et al., 1999; Green et al., 2008; Harris et al., 2008; Orem et al., 2010) and some have reported enhanced methane production with increases in surface area (Green et al., 2008), addition of inorganic nutrients (Harris et al., 2008) and trace elements (Ünal et al., 2012). The conversion of refractory OM to methane involves primarily the fermentation of polymers and monomers to fatty acids, organic acids, alcohols, hydrogen, and carbon dioxide. Subsequently degradation follows via secondary fermenting bacteria, homoacetogenic bacteria and acetoclastic, methylotrophic and hydrogenotrophic methanogens (Schink, 2006).

Similarly to coal bed formations, gas shale formations have shown enhanced methane production with increased surface area (Curtis, 2002). Horizontal drilling and hydraulic fracturing have now become two key technologies in shale gas exploration (Arthur et al., 2008; Kerr, 2010). During the drilling operation large volumes of drilling mud are pumped into the formation to cool and lubricate the drilling bit. Drilling mud contains cellulose, barite, and lignosulfonates (Caenn et al., 2011,www.fracfocus.org), which could serve as carbon and sulfate sources for microorganisms. Hydraulic fracturing is a widely used technique to fracture gas shale by pumping fluids and sand into production wells at high pressure (Arthur et al., 2008), resulting in enhanced methane extraction. Biocides are added to the fracturing water to control bacterial growth (Arthur et al., 2008), although recent studies have shown that bacteria can survive this treatment (e.g., Struchtemeyer and Elshahed, 2012) and additives such as polyacrylamide and sugar-based polymers might even be utilized microbially (Arthur et al., 2008). Despite the widespread utilization of the drilling and hydraulic fracturing procedures not much is known about the impact on the microbial communities present in formation water.

Here, we phylogenetically characterized the prokaryotic community in formation waters of three recently fractured gas-producing wells (denoted A3-11, B1-12, and C1-12) from the western margin (Manistee county) of the Antrim Shale to identify possible key microbial players in methanogenic shale gas production. Incubation experiments with formation water were performed to stimulate methanogenic communities using a variety of substrates. Direct methanogenic substrates tested include small organic acids, methanol, and trimethylamine (TMA). Sterile powdered shale, yeast extract, propionate, and glucose were used to test the ability of bacteria to convert these compounds into methanogenic substrates. Phylogenetic surveys of the incubation experiments were performed to identify the microbial communities that were stimulated by the various substrate additives. Methanol, TMA, and yeast extract substrates were monitored for consumption.

MATERIALS AND METHODS

Sampling and Sampling Location

The three investigated closely spaced (<1 km) gas-producing wells A3-11, B1-12, and C1-12 are located in the Antrim shale section of the western Michigan Basin (Manistee county) (Figure 1 and Supplementary Figure S1). All three wells were hydraulically fractured by a nitrogen foam method ~5 months prior to sampling of production water for this study. Unlike B1-12 and C1-12, A3-11 was originally drilled and brought into production more than 2 years before this study's sampling and was hydraulically fractured a total of three times, initially by different fracturing methods (Table 1). Formation water was collected at the wellhead in sterilized 1L pyrex screw cap bottles for chemical and molecular biological analyses as well as for incubation experiments. During sampling all air was removed from the collection bottles by filling them to overflowing as described previously (Huang, 2008). The anaerobic water samples were immediately placed on ice in the dark and shipped overnight to the WHOI laboratory. Well water sample bottles were transferred to an anaerobic glove box under a 95:5 N_2:H_2 atmosphere and inspected for accidental oxygen entrainment by adding well water subsamples to (clear and colorless) resazurin-containing media for color change. For DNA analyses biomass from 300-mL aliquots of native formation water from each well was collected on 0.2-μm Sterivex filters and kept frozen at −80°C until further analysis. Well water chemistry was measured with an ICP-MS system at the Analytical Chemical Testing Laboratory, Inc. (ACT, Mobile, AL).

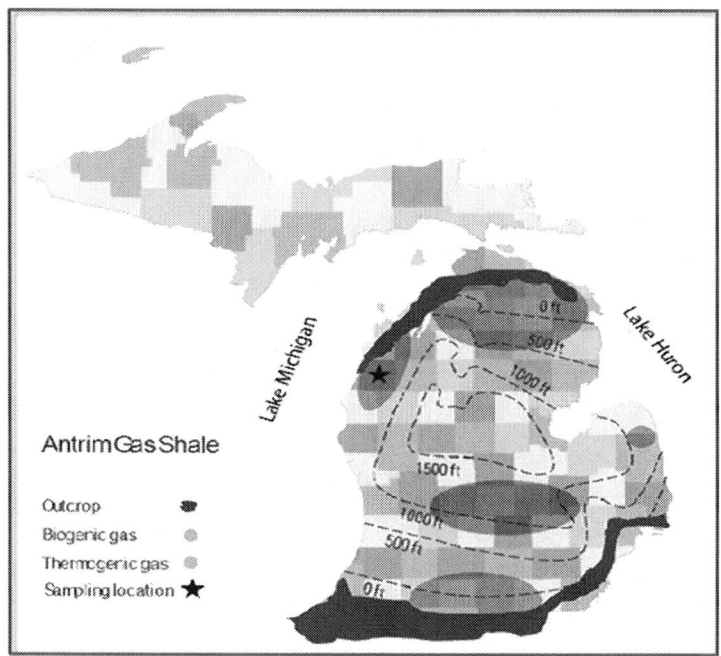

Figure 1: County map of the Antrim gas shale formation with average depth contours, outcrops, and major biogenic and thermogenic production regions delineated. The location of the wells sampled for this study is also shown.

Table 1: Chemo-physical formation water characteristics and DNA content of the studied wells

Well	Fracture history	Times fractured	pH	Alkalinity (meq. liter⁻¹)	[Cl⁻] (M)	[Na⁺] (M)	CH₄OH (mM)	[DNA] (ngL⁻¹)
A3-11	Silck water and foam fractured	3	6.97	10.6	1.72	1.26	16	1500
B1-12	Foam Fractured	1	6.56	12.5	1.51	1.13	0.1	4750
C1-12	Foam Fractured	1	6.69	13	1.73	1.29	4	1000

For more detailed information about the well water chemistry see Supplementary Tables S1, S2.

Incubation Experiments

The sampled anaerobic well water served as natural media for the incubation experiments. For the incubation experiments 45 mL of anaerobic well water was transferred to 140-mL sterile glass serum bottles and supplemented with 500 μL aliquots of sterilized anaerobic solutions of the appropriate substrate stock solutions. All steps involved in the set-up of the incubation experiment were performed in an anaerobic glove box (Coy Laboratory Products, Grass Lake, MI) under an atmosphere of at least 95% nitrogen and up to 5% hydrogen gas. Substrates included acetate, formate, glucose, propionate, and trimethylamine (TMA) at a final concentration of ~10 mM, yeast extract at a concentration of ~0.5 g/L, or a 0.44% (v/v) final concentration of methanol. Powdered sterile shale was also tested as a substrate to gauge the potential of the rock›s native bitumen as a growth substrate, with one gram added to the appropriate incubation bottles before the addition of well water. This shale material was retrieved from a 50 m core traversing the Antrim pay zone while drilling the C1-12 well. From an organic-rich section shale material was then pulverized via mortar and pestle in the lab and autoclaved dry for addition to subsequent microbial incubations.

Incubations were conducted in duplicate for each tested substrate and monitored for up to 250 days. Each bottle also received 1 mL of a 100 mM sodium sulfide solution, 500 μL of a vitamin solution, and 100 μL of a one molar phosphate solution as additional nutrients (Whitman et al., 2006) to avoid limitation by these nutrients. Following all additions, the bottles were sealed with autoclaved butyl rubber stoppers that had been boiled for 2 h in a 2 M sodium hydroxide solution to remove any traces of methane or organic contaminants. The bottles were removed from the glove box and their headspaces purged by flowing a pressurized 80:20 nitrogen:carbon dioxide mix through 0.2-μm filters and the stoppers and then over pressured to about 100 kPa, using a bench top gassing station. All bottles were incubated in the dark at room temperature (25°C) without agitation.

A 2-mL gas-tight syringe with Luer® fitting (Valco Instruments Co. Inc., Houston, TX) was used to sample headspace gases from each incubation bottle every 1–3 weeks. For each time point, 250 μL of headspace gas was withdrawn after 250 μL of sterile 80:20

nitrogen:carbon dioxide mix was pushed into the bottle from the purged syringe to avoid progressive drops in gas pressure in the bottle headspaces. A similar volume replacement procedure was used for liquid samples of 1 mL, withdrawn with sterile disposable 5-mL syringes purged with the nitrogen and carbon dioxide gas mix. Every liquid sample was divided between substrate (500 μ L filtered through a 0.2-μ m filter and frozen at −20°C) and DNA (500 μ L immediately frozen at −20°C for subsequent DNA extraction).

Microbial Gas Measurements during Incubations

The concentrations of methane and hydrogen in gas samples retrieved from each incubation bottle were quantified against a five-point calibration dilution of a custom 1 vol% CH_4/H_2 American Air Liquide standard mix on a HP 5890 Series II gas chromatograph (GC) equipped with a 2-m washed molecular sieve 13× 80/100 column and onboard thermal-conductivity and flame-ionization detectors by direct injection. Total gas abundances were then calculated from liquid and headspace volumes in each bottle and Henry's Law constants (Wilhelm et al., 1977). Anomalous readings resulting from syringe wear, identified through replicate measurements on the same or immediately preceding and succeeding sampling days, were excluded from the reported dataset. Error was propagated through these calculations using estimates of measurement error from syringes (±0.1 mL for pipetting, ±0.1 mL for 3 mL syringes, ±0.01 mL for 1 mL syringes) and standard deviations in gas concentration measurements collected from each sampling day›s standard curve measurements.

Substrate Measurements during Incubations

Standard and sample dilutions for the methanol, TMA and ethanol substrates were prepared using analytical grade reagents and distilled water that had all been sparged of background volatile organic molecules (VOMs) with helium (He) for 1 h at a rate of 100 mL/min for every 1L of solution. Analyte VOMs were likewise sparged from 1:1000 diluted liquid incubation samples using a Solatek autosampler along with a Tekmar Stratum purge and trap system outfitted with a K-type trap and quantified against a 10^{-5} to 10^{-8} calibration curve of standard dilutions on an Agilent 6850 GC (Stabilwax Crossbond Carbowax PEG capillary column, 30 m × 0.25 mm ID × 0.5 µm) (Restek Chromatography Products, Bellafonte, PA) coupled to an Agilent 5975C mass spectrometer (MS) selectively monitoring ions in scan mode of m/z 32, 43, and 45 for methanol, acetone, and isopropanol, respectively. TMA measurements were performed according to manufacturer›s recommendations employing an Rxi-624Sil capillary GC column (30 m × 0.25 mm ID × 1.4 µm) (Restek Chromatography Products, Bellafonte, PA). Briefly, 0.5 µl of 1000× diluted well fluid sample was injected directly into a 6850 Agilent GC, outfitted with a non-tapered injection port liner packed with deactivated glass wool (part# 5062-3587, Agilent Technologies, Santa Clara, CA) and programmed with a 12:1 split ratio. Detection of TMA was performed using an Agilent 5975C MS in scan mode monitoring the diagnostic ion of 58 amu.

Concentrations of total free amino acids (as a yeast extract proxy) and ammonium in 5×-diluted samples were measured at the Molecular Structure Facility of the University of California Davis using a Hitachi L-8800 amino acid analyzer.

Illumina MiSeq Sequencing of 16s rRNA Genes: Microbial Diversity and Relative Abundance in Well Water and Incubation Experiments

DNA from the initial well water and incubation experiments were extracted according to Wuchter et al. (2004) and the total DNA was quantified fluorometrically using a Quant-iT™ PicoGreen® dsDNA Reagent (Invitrogen). The V4 region of the 16S rRNA gene was amplified with universal prokaryotic primers modified from Caporaso et al. (2012). The same reverse primer 806r was used, but in combination with the 4 bp-shorter forward primer 519f (5'-CAGCMGCCGCGGTAA-3') (Øvreas et al., 1997), to increase the potential coverage of archaeal sequences (Wang and Qian, 2009). The theoretical coverage of the slightly modified primer combination was 88.1% of bacterial and 90.5% of archaeal 16S rDNA sequences available in the greengenes database (DeSantis et al., 2006) according to a primer coverage test using TestPrime 1.0 (Klindworth et al., 2013), as opposed to, respectively, 87.5 and 58.4% without modification of the forward primer. The formation of newly formed products was followed in real time using a Realplex quantitative PCR system (Eppendorf, Hauppauge, NY) and reagents (with the exception of primers) according to Coolen et al. (2009). The annealing temperature was set to 61°C and reactions were stopped in the exponential phase after 25–30 cycles. To minimize the formation of artifacts such as primer dimers, 10^8 copies were subject to a second amplification reaction with the same region-specific primers that included the Illumina flowcell adapter sequences as well as the pad regions after Caporaso et al. (2012). The reverse amplification primer now also included a unique 12 base Golay barcode sequence (Caporaso et al., 2012) to support pooling of the samples. The second qPCR run was stopped after only 12 cycles when all samples reached the end of the exponential phase. The quality of the PCR products was verified by agarose gel electrophoresis and equimolar amounts of the barcoded PCR products were pooled and purified using the AMPure®XP PCR purification kit (Agencourt Bioscience Corp., Beverly, MA). Four hundred nanogram of the mixed and purified barcoded amplicons was subject to subsequent Illumina MiSeq sequencing using the facilities of Selah Genomics (Greenville, SC).

Quality Filtering of Reads, OTU Picking, and Taxonomic Assignments

The quality scores associated with each base call for each read were used to determine the portion of each Illumina read that was of acceptable quality. Reads were first trimmed to 120 bp to avoid sequencing errors toward the end of the reads, and the minimal acceptable Phred quality score was set to 30 during demultiplexing of Fastq sequence data using the split_libraries_fastq.py script in QIIME 1.6.0 (Caporaso et al., 2010a). Furthermore, reads were discarded when they contained an N character in their sequence or barcode. The pick_otus_through_otu_table.py workflow in QIIME with default settings was used to generate an OTU table in biom format. For example, this command with default settings assigned sequences to OTUs at 97% similarity and used the RDP classifier to assign taxonomic data to each representative sequence. Then, sequences were aligned using PyNAST (Caporaso et al., 2010b) against the greengenes database and a Newick format phylogenetic tree was built for subsequent UniFrac diversity measurements. A matrix of OTU abundance in each sample with meaningful taxonomic identifiers for each OTU in biom format was assembled from the taxonomic assignments and the OTU map using default settings as the final step of the pick_otus_through_otu_ table.py workflow.

Alpha Rarefraction and Beta Diversity

Alpha rarefaction was performed in QIIME using the script make_ rarefraction_plots.py with default parameters and using the Phylogenetic Diversity, Chao1, and observed species metrics. For beta diversity analysis we executed the jackknifed_beta_diversity.py workflow script in QIIME, which used jackknife replicates to estimate the uncertainty in principal coordinate analysis (PCoA) plots and hierarchical clustering of microbial communities. This script executed the following steps: (a) Computed a beta diversity distance matrix from the full OTU table (b), produced an Unweighted Pair Group Method with Arithmetic mean (UPGMA) tree from full distance matrix, (c) and rarefied OTU tables, (d) computed distance matrices from rarefied distance matrices, (e) compared rarefied UPGMA trees and determines jackknife support for tree nodes, (f) computed principal coordinates on each rarefied distance matrix, and (g) compared rarefied PCoA plots from each rarefied distance matrix. In this step, the jackknifed replicate PCoA plots were compared to assess the degree of variation from one replicate to the next. This variation was displayed by drawing confidence ellipsoids around the samples represented in the PCoA plots.

The diversity of bacteria and archaea in the initial well waters and incubations was furthermore analyzed using domain-specific primers targeting the entire 16S rDNA-V4 region. The resulting 465-bp-long bacterial and 396 bp-long archaeal amplicons were analyzed by denaturing gradient gel electrophoresis (DGGE) (Muyzer et al., 1993) and phylogenetic analysis of capillary sequenced (trimmed to 384 bp) individual DGGE fragments. The taxonomic information based on the longer DGGE sequences was compared with the taxonomy of the numerically most abundant OTUs from the shorter Illumina reads. Methodological details of the PCR/DGGE/ capillary sequencing approach can be found in the Supplementary Info. Sequences obtained in this study were deposited to Genbank under accession numbers KC262274-KC262335 for the sequenced DGGE bands and KF72889-KF728946 for the described and discussed Illumina reads. See also supplementary information for further details.

RESULTS

Well Water Chemistry

The extant water chemistry varied slightly between the wells (Table 1). In all three wells sulfate, nitrate, and phosphate were below the detection limit (<0.10 mM). Salinities ranged between 8 and 10%. Methanol concentrations differed substantially between the three wells. The highest methanol concentration was measured in the A3-11 well water (16 mM), while methanol concentrations were four times lower in C1-12 (4 mM), and 160 times lower in B1-12 well water (~0.1 mM).

See Supplementary Tables S1, S2 in the supplementary info for detailed information about the analytical well water chemistry and gas composition. It is important to note that all three wells were likely sampled during a period of large flowback water production following hydraulic fracturing several months beforehand (see Figure 2; water data from A3-11 are not available for first 6 months of production).

Gas and water production profiles from B1-12 and C1-12 were similar, whereas those for A3-11 were an order of magnitude smaller. Unless there were much lower formation water saturations in the A3-11 area that proportionally undermined biogenic gas generation therein, such a decline in both fluid profiles relative to the two other wells insinuates a poorer completion (i.e., less well-reservoir contact creation) in A3-11. Lack of formation water dilution may in turn explain the elevated methanol concentration measured in the waters recovered from that well.

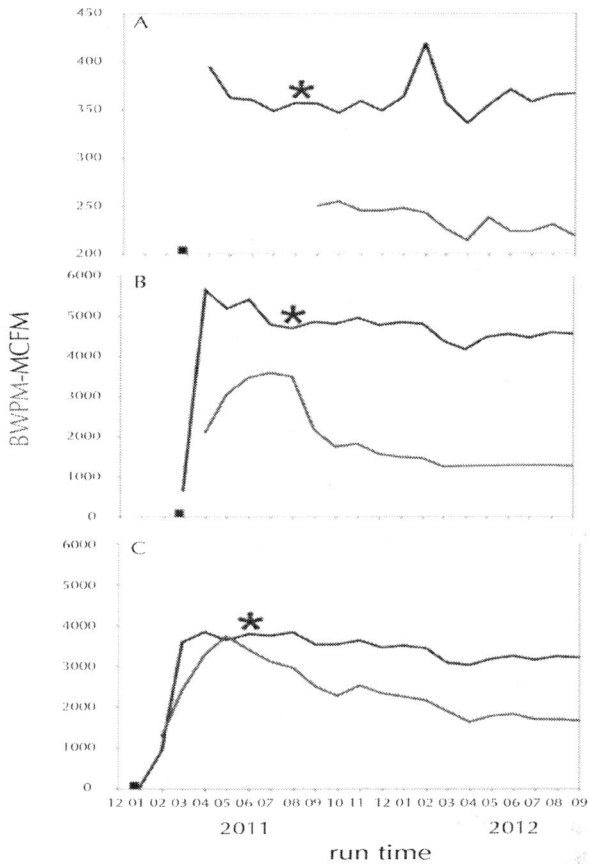

Figure 2: Monthly water and gas production in the three investigated wells. Black: Thousand cubic feet gas per month (MCFM). Blue: barrels of water per month (BWPM). Black square: time when well was fractured. Black asterisk: time when formation water was sampled for this study. (A) A3-11 well; (B) B1-12 well; (C) C1-12 well.

Prokaryotic Communities in Initial Well Waters

The DNA concentration was highest (4750 ng L^{-1}) in B1-12 well water, followed by 1500 ng L^{-1} in A3-11, and lowest (1000 ng L^{-1}) in the C1-12 well water (Table 1) and extracted DNA served as template for molecular analyses. On average 14,533 ± 6087 high quality Illumina reads with a quality score cut-off of 30 were recovered from the initial well waters.

Using a 97% sequence similarity cut-off, the total number of OTUs was twice as high in C1-12 (336 OTUs) as compared to initial well waters of A3-11 (163 OTUs) and B1-12 (164 OTUs). Rarefraction diversity measures in the initial well waters using the phylogenetic diversity metric showed that the rarefraction curves almost reached a plateau, indicating that it becomes less likely to identify new phylotypes with greater sequencing depth (Supplementary Figure S2, Supplementary Table S3). PCoA jackknifing using weighted UniFrac distance metric showed that the prokaryotic diversity was most similar between A3-11 and C1-12 (Figure 3A). The first and second axes of the PCoA analysis explained 61.9 and 16.1%, respectively, of the prokaryotic diversity variance among the three initial well waters (Figure3A).

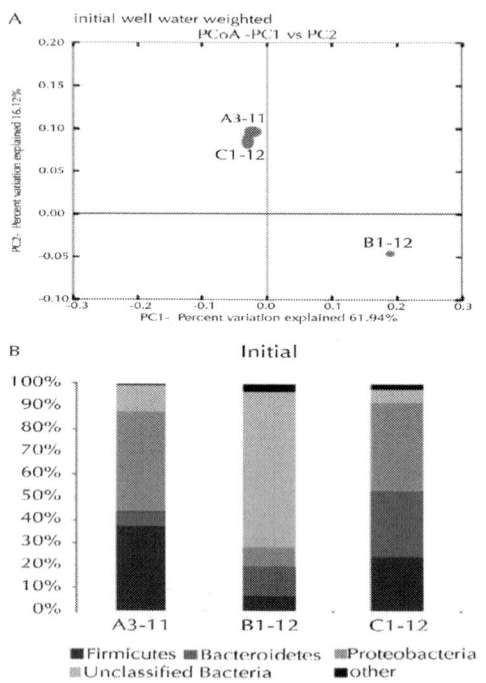

Figure 3: General overview of the prokaryotic diversity in the initial well waters. (A) Jackknifed PCoA plot of initial well water samples with weighted Unifrac. Shown is a plot of the first two principal coordinate axes, which combined explain 78% of the variation. Ellipses represent the interquartile range of the distribution of points among the 10 jackknifed replicates. (B) Relative abundance (% of Illumina reads) of the major bacterial phyla. A

more detailed overview of the most abundant unclassified bacteria is shown in Supplementary Figure S3. "Other"; euryarchaota which comprised up to 1.3% of total Illumina reads with the remainder being less abundant bacteria (mainly OP9 and Spirochaeta).

Bacterial Diversity and Relative Abundance

In all three wells >99% of the Illumina reads could be assigned with the RDP classifier in QIIME to the bacterial domain (Figure 3B). Bacteroidetes, Proteobacteria, Firmicutes and a group of unclassified bacteria comprised >96% of the Illumina reads in the studied well waters. The unclassified bacteria were most abundant in B1-12 and lowest in C1-12 (Figure 3). A single OTU with 91% similarity to the Arctic bacterium NP25 (Perreault et al., 2007) represented the majority of the unclassified bacteria, being highest (97% of the unclassified bacteria) in B1-12 (Supplementary Figure S3). This OTU represented 45% of the unclassified bacteria in C1-12, where the diversity of unclassified bacteria was highest (Supplementary Figure S3).

Bacteroidetes, Proteobacteria, and Firmicutes, which comprised more than 1% of the total reads shared 92–100% sequence similarity to previously described cultures of mesophilic and moderately halophilic bacterial species (Table 2). The majority of Bacteroidetes and Firmicutes showed highest sequence similarity with strictly anaerobic, fermenting bacteria, which can utilize a wide variety of organic compounds such as sugars, peptides, amino acids, organic acids, or alcohols. The dominantFirmicutes were most similar to described species of the order Halanaerobiales, includingHalanaerobium hydrogeniformans (Brown et al., 2011), Haloanaerobium congolense (Ravot et al., 1997), Halocella cellulolsilytica (Simankova et al., 1993), Orenia marismortui (Rainey et al., 1995),and O. salinaria (Mouné et al., 2000) (Table 2). OTUs most similar to Marinilabilia salmonicolor(Nakagawa and Yamasato, 1996) and Cytophaga sp. AN-B14 (Daffonchio et al., 2006) comprised the most abundant Bacteroidetes (Table 2). The most abundant OTUs belonging to the δ- and ε-Proteobacteria were related to sulfur-, sulfate-, iron-, or nitrate-reducing bacteria (Table 2) and a single OTU with 100% sequence similarity to the nitrate-reducing Acrobacter marinus (Kim et al., 2010) greatly outnumbered other ε-Proteobacteria as well as the δ-Proteobacteria in all three wells (Table 2). Bacterial OTUs that represented more than 1% of the Illumina reads were also recovered via capillary sequencing of individual DGGE bands (Supplementary Table S4) and the percentage sequence similarity with the most closely related cultivated species from GenBank is shown in Supplementary Table S5. However, DGGE is at best semi-quantitative and does not provide accurate information about the relative abundance of bacterial phyla and individual OTUs in the initial well waters.

Table 2: Relative abundance (% of total Illumina reads) OTUs recovered from the initial formation waters of the A3-11, B1-12, and C1-12 wells

This study OTUnr	% of reads per well			Phylum/class closest culture NCB! hit/ accession nr.	Identity (%)	Putative function
	A3-11	B1-12	C1-12			
FIRMICUTES						
1742	65	52.7	57.2	Halanaerobium hydrogenoformans IN F1_0748501	99	Fermenter
437	20.9		15.3	Haloanaerobium congolense INR_0260441	100	Fermenter, sulfur and thiosulfate reducer
1072	0	3.5	4.3	Halocella cellulolsilytica INR0369591	96	Fermenter
615	0	37.1	4.3	Orenia marismortui INF1_0262591	94	Fermenter
265	11.4	0	0	Orenia salinaria [NR _0265041	90	Fermenter
	2.7	6.7	7.2	Other		
BACTEROIDETES						
1364	27	90.7	70.7	Marinilabilia salmonicolor IAB6807211	100	Fermenter
1443	63.5	1.7	22.4	Cytophaga sp. AN-B14 1AM1576481	100	Fermenter

975	0	2.1	0	Bacteroidetes bacterium G13a-B IFN3979961	100	Acetogen
138	4	0	0	Prolixibacter bellariivorans IAB5419831	95	Fermenter
805	0	1.1	0	Bacteroidetes Phenol-4 IAF1218851	98	Phenol-degrading
667	0	0	1.1	Bacteroides graminisolvens !NR 0416421	100	Fermenter
	5.5	4.4	5.8	Other		
δ-PROTEOBACTERIA						
665	1	1.3	0	Desulfovibrio sp. AND 1 1A72813441	99	Sulfate reducer
464	0	1	0	Desulfovibrio aespoeensis (NR_0748711	100	Sulfate reducer
1362	0	1.9	21.5	Desufuromusa succinoxidans INFi_0292761	98	Sulfur reducer
1363	3.4	28.5	0	Pelobacter carbinolicus (NR_0750131	100	Iron reducer
323	0	1.2	2.9	Geoalkalibacter subterraneus IEU1822471	100	Iron reducer
99	0	12.4	0	Geobacter hephaestius IAY7375071	98	Sulfur and iron reducer
135	0	2.2	0	Geothermobacter sp. HR-1 [G01838991	98	Iron reducer
ε -PROTEOBACTERIA						
'•:s	84.7	34.1	65.1	Acrobacter marinusIEU5129201	100	Nitrate reducer
'r'. /	0	7.5	6.7	Sul furospirillum hatorespirans INR_0287711	92	Sulfur reducer
	7.2	4.6	1.2	Sulfurospirillum carboxydovorans [AY7405281	99	Sulfur reducer
	3.7	5.3	2.6	Other		

UNCLASSIFIED BACTERIA						
772	75.1	96	41.7	Arctic bacterium Np 251EU1963311	88-91	Unknown
1123	0	0	9.3	Unidentified bacteria 1F0677509!	85	Unknown
	24.9	4	49	Other		

Other reads which represent less than 1% of the reads in the major listed phyla. Other* (unclassified reads; see detailed information in Supplementary Figure S3)

Archaea Diversity and Relative Abundance

Archaeal sequences represented only 0.15, 0.46, and 1.07% of the total Illumina reads, respectively, in A3-11, B1-12, and C1-12 and are part of the "other" in Figure 3B. All detected archaeal sequences belonged to the phylum Euryarchaeota. The majority (20) of archaeal OTUs showed 83–100% sequence similarity to uncultured environmental clones, and 91% sequence similarity or less to cultured species (see Supplementary Table S4, for details). However, Illumina sequencing also revealed two archaeal OTUs with >99% sequence similarity to the methylotrophic methanogensMethanolobus profundi (Mochimaru et al., 2009) and Methanohalophilus halophilus (Wilharm et al., 1991). Two additional archaeal OTUs were most closely related to the hydrogenotrophic methanogens Methanocalculus halotolerans (Ollivier et al., 1998) and Methanoplanus limicola(Wildgruber et al., 1982). OTUs with the same most closely related GenBank sequences were also recovered from sequenced archaeal DGGE bands spanning the entire V4 region. A phylogenetic tree of the archaeal phylotypes recovered with the DGGE approach was prepared since the majority of the archaeal sequences are only distantly related to cultivated species (Supplementary Figure S4).

Incubation Experiments

Total Methane Yields in the Different Treatments

Detectable amounts of methane accumulated in the headspace of incubation bottles with all three well waters even without the addition of substrates (i.e., bottles referred to as no substrate) (Figure4). The highest stable methane yields without substrate addition were measured in incubations with A3-11 well water followed by C1-12 and being lowest in B1-12. The addition of powdered shale, acetate, and propionate did not result in increased methane yields relative to incubations without substrate addition for any well (Figure 4). The addition of glucose had a negative effect on methane generation for all three wells. Formate addition stimulated methane generation only in the B1-12 well water. The addition of yeast extract had a definite stimulatory effect on methane generation in the B1-12 and C1-12 well waters. The highest methane yields were measured with TMA and methanol as substrates with five (A3-11) to ten (B1-12) fold higher methane yields in the headspace than without substrate addition (Figure 4).

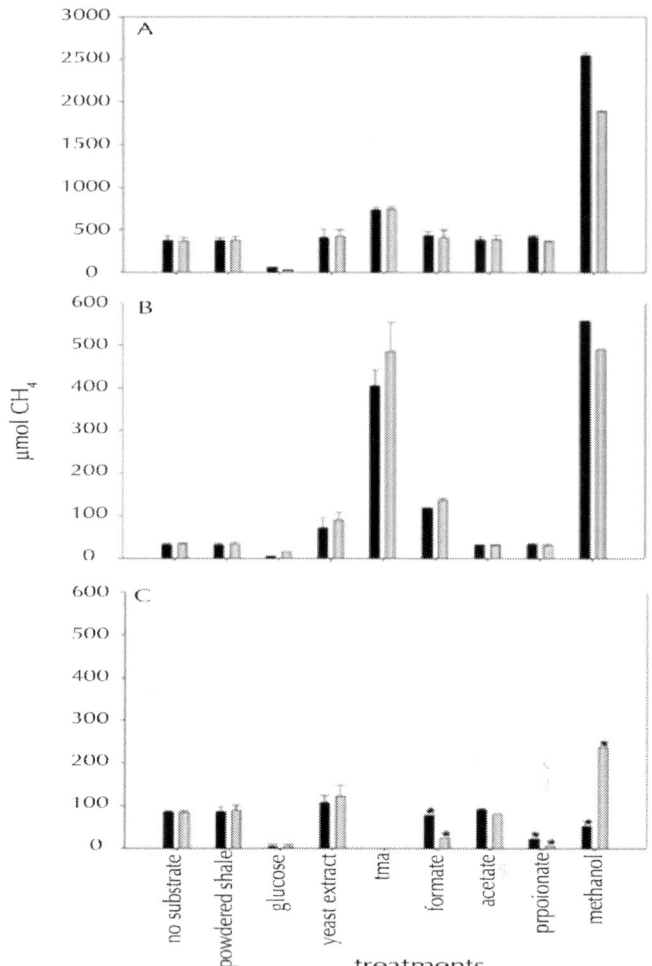

Figure 4: Average methane accumulation in formation water over the course of incubation with and without added substrates. (A) A3-11 and (B)B1-12 well waters were incubated for 202 days and (C) C1-12 well was incubated for 240 days. TMA was not tested with C1-12 well water. Note: Y-axis for (A) is 5X larger in magnitude than (B,C). *Incubation bottles where methane formation had not yet reached a plateau after 240 days of incubation. Therefore, the endpoint methane yields are shown, incubation bottle 1 (black), incubation bottle 2 (gray).

Methanol and TMA Consumption

Methanol and TMA were readily consumed by the microbial community in the incubation experiments and resulted in the highest methane yields. The utilization of methanol by methylotrophic methanogens follows according to reactions (1) and (2) as shown in Table 3. A molar methanol to methane conversion ratio [MCR_{MeOH}] of 1–1.33 would be expected (Table 3) if the methanol present in incubations with or without methanol addition were utilized completely. However, the actual measured methane accumulation in the headspace was sometimes substantially lower, suggesting that not all the methanol consumed in the no-substrate and methanol incubations was converted to methane, or that methane was consumed anaerobically in some incubation bottles (Table 4).

Table 3: Methanogenic consumption of methanol (reactions 1 and 2), TMA (reactions 3), and hydrogen (reaction 4) and molar conversion ratios

Reaction	Mola conversion ratio(MCR)
(1) $CH_3OH+H_2 \rightarrow CH_4+H_2O$	1.0
(2) $4CH_3OH \rightarrow 3CH_4+CO_2+2H_2O$	1.33
(3) $4(CH_3)3N+6H_2O \rightarrow 9CH_4+3CO_2+4NH_4+$	0.45
(4) $4H_2+CO_2 \rightarrow CH_42H_2O$	4.0

Table 4: Methanogenic substrate consumption and corresponding methane accumulation

Treatment	Methanogenic substrate	Well	Bottle	Substrate consumed in p,mol	Measured methane	MCR
No substrate	Methanol	A3-11	1	725 (.40)	407	1•8

			2	712 (.70)	341	2.1
		C1-12	1	130 (±11)	86	1.5
			2	60 (±5)	40	1.5
Methanol	Methanol	A3-11	1	3392 (±133)	1858	1.8
			2	3174 (.130)	1361	2.3
		81-12	1	1620 (±165)	418	3.9
			2	4140 (.240)	377	11.0
TMA'	TMA	43-11	1	150 (.19)	383	0.39
			2	190 (.13.5)	404	0.47
		81-12	1	181 (±5.1)	365	0.50
				. (.15.4)	409	0.35
Yeast extract	Hydrogen	B1-12	1	15	39	0.38
			2	16	45	0.36
		C1-12	1	32	22	1.45
			2	34	37	0.92
Formate'	Hydrogen	B1-12	1	211	74	2.90
			2	311	85	3.66

*Methane yields corrected for methane derived from methylotrophic methanogens after subtraction of the "no substrate" background. See Supplementary Figures S5, S6 for comparison. No MCRMeOH could be calculated for the B1-12 "no substrate" and C1-12 "methanol" experiments. The low initial methanol concentration in the B1-12 well reached the detection limit at the time of sampling in the incubation experiment. The C1-12 well water incubated with methanol addition did not reach a methane plateau at the end of the experiments (see Figure 4). Error for the methane measurement was on average between 1.1 and 2.7μmol and for hydrogen measurements 0.1 and 1.4μmol in all incubations.

TMA was tested as a methanogenic substrate for incubations with A3-11 and B1-12 well waters and resulted in methane yields comparable to those from methanol additions. TMA is utilized by methanogens according to reaction (3) as shown in Table 3. Given the stoichiometry of reaction (3), complete TMA consumption would theoretically result in a TMA to methane conversion ratio [MCR_{TMA}] of 0.45. The measured methane yields, when corrected for methanol-derived methane productions, are in good agreement with theoretical expectations suggesting that most consumed TMA was converted to methane. See Supplementary Figure S5 for detailed information about the methanol and TMA consumption and methane accumulation in the A3-11 well over the course of incubation.

Hydrogen Dynamics in the Different Treatments

Gradual accumulation of hydrogen in excess of 200 µmoles was detected in the headspace of all three well waters when incubated with glucose and formate. Only formate addition to B1-12 well water resulted in hydrogen consumption and concomitant methane production (Table 4). No hydrogen consumption was observed in well waters incubated with glucose, and methane accumulation was minor in all three well waters over the entire course of incubation (Figures 4 and Supplementary Figure S6). A lower amount of hydrogen (15–34 µmoles) was detected in the headspace of well water incubated with yeast-extract (Table 4) (yeast extract consumption; Supplementary Figures S6, S7). Incubation with the other substrates did not result in detectable accumulation of hydrogen.

Hydrogenotrophic methanogens utilize hydrogen according to reaction (4) as shown in Table 3. According to this stoichiometry, for every mole of methane produced, four moles of hydrogen must be consumed. When corrected for background methane production (as measured from the no-substrate bottles), for example, in the C1-12 well between 22 and 37 µmoles of methane was detected in well waters incubated with yeast extract. Such a methane yield would correspond to a range of 88 to 148 µmoles of hydrogen consumption. This is far greater than the measured hydrogen accumulations (Table 4), suggesting that part of the produced hydrogen escaped detection due to quick turnover by hydrogenotrophic methanogens in the incubations.

Effect of Surface Area

Methane accumulated substantially earlier in incubations with powdered shale compared to well waters without substrate addition (data only shown for C1-12 well, Figure 5). Despite the difference in initial methane production rates, the amount of methane production with or without the addition of shale had stabilized at nearly identical methane yields toward the end of the incubation period. Approximately 25 µmoles of methane that was produced in C1-12 well water with sterile pulverized shale might have been consumed microbially as evident from the difference in the amount of methane after 30 and 240 days of incubation (Figure 5).

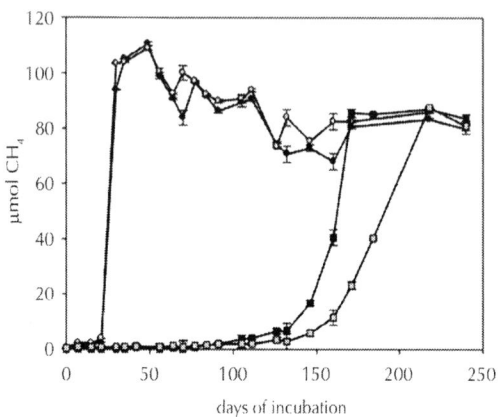

Figure 5: Methane development in the C1-12 well water with sterile powdered shale as substrate (bottle 1, bottle 2), and no substrate treatment (bottle 1 and bottle 2).

Prokaryotic Community Detected in the Incubation Experiments

Bacteria

Most bacterial OTUs found in the incubation experiments were also present in the initial well water (Supplementary Figure S8). In the incubation experiments the dominant Firmicutes were OTUs with highest sequence similarity to Halanerobium hydrogeniformans, H. congolense, Orenia marismortui, and O. salinaria. OTUs with highest sequence similarity to Marinilabilia salmonicolor and Prolixibacter bellariivorans comprised the most abundant Bacteroidetes. TheProteobacteria were predominated by OTUs with highest sequence similarity to Acrobacter marinus, Pelobacter carbinolicus, Sulfurospirillum sp., and Desulfovibrio sp. (Supplementary Figure S8). However, the relative abundance of these microbial populations was found to differ significantly between treatments (ANOSIM; P = 0.024; r = 0.2658), which is further visualized by the limited overlap between samples in the Jacknifed PCoA plot (Supplementary Figure S9).

In all three well waters the addition of phosphate and nutrients stimulated the growth of bacteria belonging to the Firmicutes, reaching > 80% of the total bacterial Illumina reads (Supplementary Figure S8). Powdered shale had the largest effect on the microbial communities in C1-12 with a shift toward a predominance of Proteobacteria as compared to the incubations without substrate (Supplementary Figure S8). The addition of methanol stimulated Firmicutes in A3-11 andProteobacteria in the two other well waters compared to no substrate addition. Glucose had little effect on bacterial communities in A3-11 and C-1-12, but stimulated Proteobacteria in B1-12. Yeast extract, TMA and formate stimulated the growth of unclassified bacteria, notably a bacterium with highest sequence similarity to the uncultured Arctic bacterium NP25, in the majority of well water incubations. Acetate and propionate stimulated the growth of Proteobacteria in B1-12 and C1-12, while the relative abundance of uncultured bacteria was stimulated in A3-11 upon the addition of acetate (Supplementary Figure S8).

Archaea

OTUswith>99%sequencesimilaritytothemethylotrophicmethanogens Methanolobus profundiand Methanohalophilus halophilus, were detected in all incubation experiments with all tested substrates. Sequences with 99% sequence similarity to the hydrogenotrophic methanogenMethanoplanus limicola were found in the yeast extract treatment in B1-12 and C1-12 well waters. Sequences with 100% similarity to another hydrogenotrophic methanogen, Methanocalculus halotolerans, were found only in formate-treated B1-12 well water. Archaeal OTUs detected in the initial well water without close cultured representatives were not detected in the incubation experiments with one exception; archaeal OTU O5 was found in methanol-treated B1-12 well water (Supplementary Table S5). Despite methane formation and the presence of OTUs related to methanogenic archaea, the relative abundance of archaeal Illumina reads did not increase in any of the incubations compared to the initial.

DISCUSSION

Production Water Characteristics and Implication on Microbial Community

The methane-producing zone in the Antrim Shale formation contain formation waters whose temperature and salinity range is suitable for mesophilic moderately halophilic microorganisms (Martini et al., 1996; Waldron et al., 2007). Nonetheless, as mentioned earlier, the three wells were all recently fractured and their production water profiles suggest that water samples collected for this study might have entrained volumetrically significant amounts of hydraulic fracture flowback fluid, with an especially large proportional contribution implied for the A3-11 well. Thus, the well water microbial community identified in our survey most likely represents a mixture of indigenous shale communities and allochthonous species, which were introduced to these reservoirs during drilling and fracturing procedures (Struchtemeyer et al., 2011; Struchtemeyer and Elshahed, 2012). The residence time of the introduced drilling mud and fracturing fluid into the shale formation impacts the formation water chemistry and biology. During the drilling procedure large volumes of drilling mud are often lost in the shale formation (Gray et al., 1980; Grace, 2007) and only 30–70% of fracturing fluids injected into wells are recovered in the flowback waters (Veil, 2010). A recent study on hydraulic fractured thermogenic wells in the Barnett Shale (USA) showed that the concentration of salt, iron and total dissolved solids were higher in flowback water of a well that had been in contact with fracturing fluids for 2 months compared to a well with a much shorter contact time of 24 h (Struchtemeyer and Elshahed, 2012). The authors suggested that the differences in the bacterial community in the flowback water of the two wells might be influenced by the different time intervals (2 month vs. 24 h) between fracturing and flowback at the two sites (Struchtemeyer and Elshahed, 2012). Our data support these findings. Namely, the prokaryotic beta diversity in the two wells (A3-11 and C1-12) with a longer contact time of fracturing fluid was more similar than in B1-12 with a shorter contact time of fracturing fluid as indicated by the differences in methanol concentration. In addition to methanol, cellulose, lignosulfonates, and sugar-based polymers are components of fracturing fluids and drilling muds (Caenn et al., 2011, www.fracfocus.org). For example, guar gum or hydroxyethyl cellulose is frequently used in fracturing fluids to thicken the water in order to suspend the sand. These additives might explain the presence of an active fermenting microbial community in the production water as observed in the incubation experiments by the quick utilization of fermentable substrates such as yeast extract and glucose. The prominent heterotrophs in the production water and the incubation experiments were members of the genusHaloanaerobium (Firmicutes) and the genera Marinilabilia and Cytophaga (Bacteroidetes). Haloanaerobium species usually ferment saccharides and produce H_2, CO_2, and C_2 compounds (mainly acetate and sometimes ethanol, as was detected in the glucose treated incubations) (Ollivier and Cayol, 2005). One prominent OTU was closely related to Marinilabilia salmonicolor (Nakagawa and Yamasato, 1996), a bacterium capable of utilizing cellulose and complex carbohydrate. Additional sequences were related to Cytophaga, Dechloromonas, and Pseudomonas and members of these genera have been reported to be capable of growing on crude oil, benzene, xylene, and toluene (Prince, 2005), and might play an important role in the degradation of refractory OC in the Antrim shale. Also, the unclassified bacteria constitute a substantial fraction of the total bacterial reads, especially in the B1-12 well. These bacteria might also be capable of utilizing complex organic shale matter. Furthermore, a number of archaeal OTUs unaffiliated with known methanogenic groups were detected in the initial production water samples, which were not subsequently detected during the bottle incubations. These euryarchaeota may possess non-methanogenic metabolisms and also be involved in bitumen degradation in situ.

Methanogenic Capabilities in Incubation Experiments

Methanol was quickly utilized in the incubation experiments and high concentrations of added methanol did not inhibit methylotrophic methanogenesis suggesting that the methanogenic community is well-adapted to high methanol concentrations most likely derived from the fracturing procedure since methanol is a standard ingredient in fracturing fluids. Halophilic methanogens generally utilize substrates such as methanol, methylamines, or dimethyl sulfide, which yields more free energy compared to acetate or hydrogen (Oren, 2001). This extra energy might be used for their osmoregulatory system to balance the energetic demands of the saline environment (Oren, 2001).

TMA was below detection limit in our formation waters suggesting that this compound plays a minor role in comparison to methanol as a substrate for methylotrophic methanogens in this environment. TMA could potentially be formed in formation water through fermentation of betaine, which is an osmoregulatory compound synthesized by some halophilic bacteria (Oren, 1990). The bacterial fermentation of betaine can yield acetate and TMA. Also, similar to coal matrices (Strąpoć et al., 2011) the organic-rich Antrim Shale might be a source of methylamines.

Acetate was not utilized by methanogens in any of the three analyzed well waters and we did not find any phylogenetic evidence for acetoclastic methanogens. This is in agreement with the finding that acetoclastic and hydrogenotrophic methanogens generally thrive in freshwater or lower salinity environments (Oren, 1999, 2001). Instead, hydrogenotrophic methanogens were detected in the initial well waters and after incubation with yeast extract. Hydrogenotrophic methanogens were also identified when formate was used as a substrate, but only in B1-12. Here, hydrogen was generated by fermentation and consumed concomitant with methane production. B1-12 incubated with formate yielded an OTU closely related to Methanocalculus halotolerans, the most halotolerant hydrogenotrophic methanogen known to date and capable of withstanding concentrations of up to 12% NaCl (Ollivier et al., 1998). One OTU closely related to Methanoplanus limicola (Wildgruber et al., 1982) was detected in B1-12 and C1-12 well waters incubated with yeast extract. This methanogen tolerates salt concentrations between 0.4 and 5.4% (Wildgruber et al., 1982), which is substantially lower than the salt concentration of 8–10% in our formation waters. Most likely this OTU represents a novel hydrogenotrophic methanogen with a higher salinity tolerance. Both hydrogenotrophic methanogens might possess different hydrogen affinities since the utilization of formate resulted in a 10-fold higher hydrogen concentration then the amount of hydrogen resulting from the fermentation of yeast extract. The OTU related to M. limicola, might be better adapted to grow at low hydrogen concentrations, whereas the OTU related to M. halotolerans, is capable of thriving at elevated hydrogen concentrations. This could be similar to methanogens from rice paddies, which grow slowly if at all at high hydrogen concentrations (Sakai et al., 2007).

The low methane generation in the glucose-treated incubations could be explained by an inhibition of the methanogenic community caused by the rapid depletion of nutrients and/or the decrease in pH from the accumulation of organic acid intermediates (Str po et al., 2011) as a result of bacterial fermentation. The accumulation of ethanol caused by fermentation might also have had an inhibitory effect on the methanogens present in the glucose treatment.

The increase of mineral surface area greatly stimulated methane production rates as observed from formation waters that were incubated with sterile powdered shale. However, the addition of shale did not result in consistent increases in methane yield in the C1-12 well water, and the overall methane yield in the other two well waters was identical to the no substrate background. This suggests that over the course of incubation the bitumen in the powdered shale was not utilized by fermentative prokaryotes and did not yield substrates for the methanogenic communities. In contrast, a previous enrichment experiment showed that fermentative bacteria derived from Antrim well water could be enriched by using only water-soluble shale OM and subsequent methane accumulation was detected, demonstrating that methanogenic communities can be supported by using shale derived DOM as the only source of energy and carbon (Huang, 2008). Most likely our fermenting microbial community was adapted to the presence of easy degradable carbon compounds such as those stemming from the drilling and fracturing procedure, and was not adapted to degrade bitumen from the shale.

Methanol and Methane Consumption

The discrepancy between the theoretically expected methane yields based on methanol consumption and measured methane yields requires non-methanogenic methanol-utilizing metabolisms, or the ability of anaerobic oxidation of methane to be present. This could be explained by OTUs related to methanol-utilizing bacteria such as Geoalkalibacter subterraneus (Greene et al., 2009) present in the incubation experiments. Furthermore, some sulfate reducers are known to utilize methanol (Qatibi et al., 1991; Tarasov et al., 2011), and might be involved in methanol consumption. Also, the detected archaea without close cultured relatives, which were detected in the methanol incubation experiment, might possess the capability to utilize methanol.

The depletion of methane suggests anaerobic methane oxidation (AMO) at least in C1-12 well water incubated with powdered shale. Methane can be oxidized anaerobically by a consortium of methane-oxidizing archaea (ANME clusters), and sulfate-reducing bacteria (SRB) (Knittel and Boetius, 2009). We did find sequences of SRB in our incubation experiments, however no ANME-related sequences were recovered despite a good coverage of the universal primers used to generate template DNA for Illumina sequencing. Although the archaeal partner involved in AMO was not found, the depletion of methane in some of the incubations and possible involvement of AMO is in agreement with previous geochemical evidence for anaerobic hydrocarbon oxidation in formation water of the western margin of the Antrim shale (Martini et al., 2003). The western margin formation water contains some of the highest sulfate concentrations relative to salinity in the Antrim shale (Walter et al., 1997), and C2 and C3 gasses are enriched in ^{13}C values, suggesting that anaerobe hydrocarbon oxidation occurred in the western margin (Martini et al., 2003).

The sulfate present in the formation water is most likely derived from the shale since a recent incubation-based study showed that water-soluble organic shale material sustains SRB (Huang, 2008). This is supported by our incubation experiments where a shift toward - and -Proteobacteria, which comprise many iron-, sulfur-, sulfate-, and nitrate reducers, was observed with the addition of powdered shale.

Sulfate can be quickly utilized by SRB, which compete for hydrogen and other low-molecular-weight substrates with methanogens. Recently, a study on formation water from wells along the northern production trend of the Antrim shale showed an increase in sulfate concentrations and SRB over the last two decades (Kirk et al., 2012). The authors explained these changes by ongoing processes driven by commercial gas production such as ground water inflow from the sulfate-rich Travers limestone, which is in contact with the Antrim shale (Kirk et al., 2012). It was furthermore suggested that this development could have negative implications for commercial gas production by creating conditions that favor growth of SRB (Kirk et al., 2012). These bacteria can compete with methanogens for substrate or, as indicated previously (Martini et al., 2003) and suggested in our study, SRB could be involved in AMO and responsible for loss of methane in the shale formation.

CONCLUDING REMARKS

Microbial community composition in closely spaced production wells can differ substantially due to local variation in both reservoir quality (e.g., lithology, matrix and natural fracture microstructure, fluid chemistry and saturation) and completion quality (e.g., hydraulic fracture design and fluid composition, well cleanup workflow, and subsequent production strategy). The microbial fermenting community in our study was capable of rapidly utilizing substrates, such as glucose and yeast extract, suggesting that they are well-adapted to decompose relatively labile OM including organic additives used in drilling and fracturing fluids. Hydrogenotrophic methanogens were detected, but the production waters from recently fractured wells appear to be dominated by methylotrophic methanogens, capable of utilizing high concentrations of methanol likely stemming from the fracturing fluids. Furthermore, we found that increased surface area stimulated methane production. However, a loss of methane over the course of incubation suggests that AMO also occurs in formation waters, which might constitute a substantial loss of methane in the shale formation.

ACKNOWLEDGMENTS

This study was conducted at Woods Hole Oceanographic Institution (WHOI) and Schlumberger Doll Research (SDR) with support from the Schlumberger Future of Research (FOR) proposal program and SDR Department of Reservoir Geosciences. We are grateful to Sean Silva and Carl Johnson for technical assistance with gas chromatography and to Dr. Chris Reddy for providing the anaerobic glove box at WHOI. We also thank John Miller, Linda Hasbrouck, and their colleagues at Presidium Antrim West LC who facilitated the collection and shipping of the water and core samples used in this study. Tracy J. Mincer was supported for this project from NSF CHE-OCE Award # 1131415.

REFERENCES

1.   Arthur, J., Bohm, B., Coughlin, B. J., and Layne, M. (2008). Evaluating the Environmental Implications of Hydraulic Fracturing in Shale Gas Reservoirs. Tusla: ALL Consulting.

2.   Brown, S. D., Begemann, M. B., Mormile, M. R., Wall, J. D., Han, C. S., Goodwin, L. A., and et al. (2011). Complete genome sequence of the haloalkaliphilic, hydrogen-producing bacterium Halanaerobium hydrogeniformans. J. Bacteriol. 193, 3682–3683. doi: 10.1128/JB.05209-11

3.   Caenn, R., Darley, H. C. H., and Gray, G. R. (2011). Composition and Properties of Drilling and Completion Fluids, 6th Edn. Oxford: Elsevier.

4.   Caporaso, J. G., Kuczynski, J., Stombaugh, J., Bittinger, K., Bushman, F. D., Costello, E. K., and et al. (2010a). QIIME allows analysis of high-throughput community sequencing data. Nat. Methods 7, 335–336. doi: 10.1038/nmeth.f.303

5.   Caporaso, J. G., Bittinger, K., Bushman, F. D., DeSantis, T. Z., Andersen, G. L., and Knight, R. (2010b). PyNAST: a flexible tool for aligning sequences to a template alignment. Bioinformatics 26, 266–267. doi: 10.1093/bioinformatics/btp636

6.   Caporaso, J. G., Lauber, C. L., Walters, W. A., Berg-Lyons, D., Huntley, J., Fierer, N., and et al. (2012). Ultra-high-throughput

microbial community analysis on the Illumina HiSeq and MiSeq platforms. ISME J. 6, 1621–1624. doi: 10.1038/ismej.2012.8

7. Coolen, M. J. L., Saenz, J. P., Giosan, L., Trowbridge, N. Y., Dimitrov, P., Dimitrov, D., and et al. (2009). DNA and lipid molecular stratigraphic records of haptophyte succession in the Black Sea during the Holocene. Earth Planet. Sci. Lett. 284, 610–621. doi: 10.1016/j.epsl.2009.05.029

8. Curtis, J. B. (2002). Fractured shale -gas systems. Am. Assoc. Pet. Geol. Bull. 86, 1921–1938. doi: 10.1306/61EEDDBE-173E-11D7-8645000102C1865D

9. Daffonchio, D., Borin, S., Brusa, T., Brusetti, L., van der Wielen, P. W. J. J., Bolhuis, H., and et al. (2006). Stratified prokaryote network in the oxic-anoxic transition of a deep-sea halocline. Nature 440, 203–207. doi: 10.1038/nature04418

10. DEQ. (2012). Department of Environmental Quality, Michigan. Office of Oil, Gas and Minerals. Available online at:http://www.michigan.gov/documents/deq/MICHIGAN_OIL_GAS_MAP_411600_7.pdf

11. DeSantis, T. Z., Hugenholtz, P., Larsen, N., Rojas, M., Brodie, E. L., Keller, K., and et al. (2006). Greengenes, a chimera-checked 16S rRNA gene database and workbench compatible with ARB. Appl. Environ. Microbiol. 72, 5069–5072. doi: 10.1128/AEM.03006-05

12. Grace, R. (2007). Oil: an Overview of the Petroleum Industry. Houston, TX: Gulf Publishing Co.

13. Gray, G. R., Darley, H. C. H., and Rogers, W. F. (1980). Composition and Properties of Oil Well Drilling Fluids. Houston, TX: Gulf Publishing Co.

14. Green, M. S., Flanegan, K. C., and Gilcrease, P. C. (2008). Characterization of a methanogenic consortium enriched from a coalbed methane well in the Powder River Basin, USA. Int. J. Coal Geol. 76, 34–45. doi: 10.1016/j.coal.2008.05.001

15. Greene, A. C., Patel, B. K. C., and Yacob, S. (2009). Geoalkalibacter subterraneus sp nov., an anaerobic Fe(III)- and Mn(IV)-reducing bacterium from a petroleum reservoir, and emended descriptions of the family Desulfuromonadaceae and the genus Geoalkalibacter. Int. J. Syst. Evol. Microbiol. 59, 781–785. doi: 10.1099/ijs.0.001537-0

16.   Harris, S. H., Smith, R. L., and Barker, C. E. (2008). Microbial and chemical factors influencing methane production in laboratory incubations of low-rank subsurface coals. Int. J. Coal Geol. 76, 46–51. doi: 10.1016/j.coal.2008.05.019

17.   Huang, R. (2008). Shale-Derived Dissolved Organic Matter as a Substrate for Subsurface Methanogenetic Communities in the Antrim Shale Michigan Basin, USA. Master thesis. Department of Geosciences, University of Massachusetts Amherst.

18.   Kerr, R. A. (2010). Natural gas from shale bursts onto the scene. Science 328, 1624–1626. doi: 10.1126/science.328.5986.1624

19.   Kim, H. M., Hwang, C. Y., and Cho, B. C. (2010). Arcobacter marinus sp nov. Int. J. Syst. Evol. Microbiol. 60, 531–536. doi: 10.1099/ijs.0.007740-0

20.   Kirk, M. F., Martini, A. M., Breecker, D. O., Colman, D. R., Takacs-Vesbach, C., and Petsch, S. T. (2012). Impact of commercial natural gas production on geochemistry and microbiology in shal-gas reservoir. Chem. Geol. 332–333, 15–25. doi: 10.1016/j.chemgeo.2012.08.032

21.   Klindworth, A., Pruesse, E., Schweer, T., Peplies, J., Quast, C., Horn, M., and et al. (2013). Evaluation of general 16S ribosomal RNA gene PCR primers for classical and next-generation sequencing-based diversity studies. Nucleic Acids Res. 41, e1. doi: 10.1093/nar/gks808

22.   Knittel, K., and Boetius, A. (2009). Anaerobic oxidation of methane: progress with an unknown process. Annu. Rev. Microbiol. 63, 311–334. doi: 10.1146/annurev.micro.61.080706.093130

23.   Magot, M., Ollivier, B., and Patel, B. K. C. (2000). Microbiology of petroleum reservoirs. Antonie Van Leeuwenhoek 77, 103–116. doi: 10.1023/A:1002434330514

24.   Martini, A. M., Budai, J. M., Walter, L. M., and Schoell, M. (1996). Microbial generation of economic accumulations of methane within a shallow organic-rich shale. Nature 383, 155–158. doi: 10.1038/383155a0

25.   Martini, A. M., Nüsslein, K., and Petsch, S. T. (2004). Enhancing Microbial Gas from Unconventional Reservoirs: Geochemical and Microbiological Characterization of Methane-Rich Fractured Black Shales. Final Report: RPSEA-0024-04, GRI-05/0023. Gas Technology Institute, Des Plaines, IL.

26. Martini, A. M., Walter, L. M., Budai, J. M., Ku, T. C. W., Kaiser, C. J., and Schoell, M. (1998). Genetic and temporal relations between formation waters and biogenic methane: Upper Devonian Antrim Shale, Michigan Basin, USA. Geochim. Cosmochim. Acta 62, 1699–1720. doi: 10.1016/S0016-7037(98)00090-8

27. Martini, A. M., Walter, L. M., Ku, T. C. W., Budai, J. M., McIntosh, J. C., and Schoell, M. (2003). Microbial production and modification of gases in sedimentary basins: a geochemical case study from a Devonian shale gas play, Michigan basin.Am. Assoc. Pet. Geol. Bull. 87, 1355–1375. doi: 10.1306/031903200184

28. Martini, A. M., Walter, L. M., and McIntosh, J. C. (2008). Identification of microbial and thermogenic gas components from Upper Devonian black shale cores, Illinois and Michigan basins. Am. Assoc. Pet. Geol. Bull. 92, 327–339. doi: 10.1306/10180706037

29. McIntosh, J. C., Walter, L. M., and Martini, A. M. (2002). Pleistocene recharge to midcontinent basins: effects on salinity structure and microbial gas generation. Geochim. Cosmochim. Acta 66, 1681–1700. doi: 10.1016/S0016-7037(01)00885-7

30. Milkov, A. V. (2011). Worldwide distribution and significance of secondary microbial methane formed during petroleum biodegradation in conventional reservoirs. Org. Geochem. 42, 184–207. doi: 10.1016/j.orggeochem.2010.12.003

31. Mochimaru, H., Tamaki, H., Hanada, S., Imachi, H., Nakamura, K., Sakata, S., and et al. (2009). Methanolobus profundisp nov., a methylotrophic methanogen isolated from deep subsurface sediments in a natural gas field. Int. J. Syst. Evol. Microbiol. 59, 714–718. doi: 10.1099/ijs.0.001677-0

32. Mouné, S., Eatock, C., Matheron, R., Willison, J. C., Hirschler, A., Herbert, R., and et al. (2000). Orenia salinaria sp. nov., a fermentative bacterium isolated from anaerobic sediments of Mediterranean salterns. Int. J. Syst. Evol. Microbiol. 50, 721–729. doi: 10.1099/00207713-50-2-721

33. Muyzer, G., De Waal, E., and Uitterlinden, A. (1993). Profiling of complex microbialpopulations by denaturing gradient gelelectrophoresis analysis of polymerase chain reaction-amplified genes coding for 16s rRNA. Appl. Environ. Microbiol. 59, 695–700.

34.   Nakagawa, Y., and Yamasato, K. (1996). Emendation of the genus Cytophaga and transfer of Cytophaga agarovorans andCytophaga salmonicolor to Marinilabilia gen nov: phylogenetic analysis of the Flavobacterium-Cytophaga complex. Int. J. Syst. Bacteriol. 46, 599–603. doi: 10.1099/00207713-46-2-599

35.   Ollivier, B., and Cayol, J. L. (2005). "Fermentative, iron-reducing and nitrate-reducing microorganisms," in Petroleum Microbiology, eds B. Ollivier and M. Magot (Washington, DC: ASM press), 71–88.

36.   Ollivier, B., Fardeau, M. L., Cayol, J. L., Magot, M., Patel, B. K. C., Prensier, G., and et al. (1998). Methanocalculus halotolerans gen. nov., sp. nov., isolated from an oil-producing well. Int. J. Syst. Bacteriol. 48, 821–828. doi: 10.1099/00207713-48-3-821

37.   Orem, W. H., Voytek, M. A., Jones, E. J., Lerch, H. E., Bates, A. L., Corum, M. D., and et al. (2010). Organic intermediates in the anaerobic biodegradation of coal to methane under laboratory conditions. Org. Geochem. 41, 997–1000. doi: 10.1016/j.orggeochem.2010.03.005

38.   Oren, A. (1990). Formation and breakdown of glycine betaine and trimethylamine in hypersaline environments. Antonie Van Leeuwenhoek 58, 291–298. doi: 10.1007/BF00399342

39.   Oren, A. (1999). Bioenergetic aspects of halophilism. Microbiol. Mol. Biol. Rev. 63, 334.

40.   Oren, A. (2001). The bioenergetic basis for the decrease in metabolic diversity at increasing salt concentrations: implications for the functioning of salt lake ecosystems. Hydrobiologia 466, 61–72. doi: 10.1023/A:1014557116838

41.   Øvreas, L., Forney, L., Daae, F., and Torsvik, V. (1997). Distribution of bacterioplankton in meromictic lake Saelenvannet, as determined by denaturing gradient gel electrophoresis of PCR-amplified gene fragments coding for 16S rRNA. Appl. Environ. Microbiol. 63, 3367–3373.

42.   Perreault, N. N., Andersen, D. T., Pollard, W. H., Greer, C. W., and Whyte, L. G. (2007). Characterization of the prokaryotic diversity in cold saline perennial springs of the Canadian high Arctic. Appl. Environ. Microbiol. 73, 1532–1543. doi: 10.1128/AEM.01729-06

43. Prince, R. C. (2005). "The microbiology of marine oil spill bioremediation," in Petroleum Microbiology, eds B. Ollivier and M. Magot (Washington, DC: ASM press), 317–336.

44. Qatibi, A. I., Cayol, J. L., and Garcia, J. L. (1991). Glycerol and propanediols degradation by Desulfovibrio-Alcoholovoransin pure culture in the presence of sulfate, or in syntrophic association with Methanospirillum-Hungatei. FEMS Microbiol. Ecol. 85, 233–240. doi: 10.1111/j.1574-6941.1991.tb01728.x

45. Rainey, F. A., Zhilina, T. N., Boulygina, E. S., Stackebrandt, E., Tourova, T. P., and Zavarzin, G. A. (1995). The taxonomic status of the fermentative halophilic anaerobic-bacteria - description of Haloanaerobiales ord-nov, Halobacteroidaceae fam-nov, Orenia gen-nov and further taxonomic rearrangements at the genus and species level. Anaerobe 1, 185–199. doi: 10.1006/anae.1995.1018

46. Ravot, G., Magot, M., Ollivier, B., Patel, B. K. C., Ageron, E., Grimont, P. A. D., and et al. (1997). Haloanaerobium congolense sp nov, an anaerobic, moderately halophilic, thiosulfate- and sulfur-reducing bacterium from an African oil field.FEMS Microbiol. Lett. 147, 81–88. doi: 10.1111/j.1574-6968.1997.tb10224.x

47. Rice, D. D. (1992). "Controls, habitat, resource potential of ancient bacterial gas," in Bacterial Gas, ed R. Vitaly (Paris: Editions Technip), 91–118.

48. Sakai, S., Imachi, H., Sekiguchi, Y., Ohashi, A., Harada, H., and Kamagata, Y. (2007). Isolation of key methanogens for global methane emission from rice paddy fields: a novel isolate affiliated with the clone cluster Rice Cluster, I. Appl. Environ. Microbiol. 73, 4326–4331. doi: 10.1128/AEM.03008-06

49. Schink, B. (2006). Microbially driven redox reactions in anoxic environments: Pathways, energetics, and biochemical consequences. Eng. Life Sci. 6, 228–233. doi: 10.1002/elsc.200620130

50. Shumkov, S., Terekhova, S., and Laurinavichius, K. (1999). Effect of enclosing rocks and aeration on methanogenesis from coals. Appl. Microbiol. Biotechnol. 52, 99–103. doi: 10.1007/s002530051494

51. Shurr, G. W., and Ridgley, J. L. (2002). Unconventional shallow biogenic gas systems. Am. Assoc. Pet. Geol. Bull. 86, 1939–1969. doi: 10.1306/61EEDDC8-173E-11D7-8645000102C1865D

52. Simankova, M. V., Chernych, N. A., Osipov, G. A., and Zavarzin, G. A. (1993). Halocella Cellulolytica gen-nov, sp-nov, a new obligately anaerobic, halophilic, cellulolytic bacterium. Syst. Appl. Microbiol. 16, 385–389. doi: 10.1016/S0723-2020(11)80270-5

53. Strąpoć, D., Picardal, F. W., Turich, C., Schaperdoth, I., Macalady, J. L., Lipp, J. S., and et al. (2008). Methane-producing microbial community in a coal bed of the Illinois basin. Appl. Environ. Microbiol. 74, 3918–3918. doi: 10.1128/AEM.00856-08

54. Strąpoć, D., Mastalerz, M., Dawson, K., Macalady, J., Callaghan, A. V., Wawrik, B., and et al. (2011). Biogeochemistry of Microbial Coal-Bed Methane. Annu. Rev. Earth Planet. Sci. 39, 617–656. doi: 10.1146/annurev-earth-040610-133343

55. Struchtemeyer, C. G., Davis, J. P., and Elshahed, M. S. (2011). Influence of drilling mud formulation process on the bacterial communities in thermogenic natural wells of the Barnett Shale. Appl. Environ. Microbiol. 77, 4744–4753. doi: 10.1128/AEM.00233-11

56. Struchtemeyer, C. G., and Elshahed, M. S. (2012). Bacterial communities associated with hydraulic fracturing fluids in thermogenic natural gas wells in North Central Texas, USA. FEMS Microbiol. Ecol. 81, 13–25. doi: 10.1111/j.1574-6941.2011.01196.x

57. Tarasov, A. L., Borzenkov, I. A., and Belyayev, S. S. (2011). Investigation of the trophic relations between anaerobic microorganisms from an underground gas repository during methanol utilization. Microbiology 80, 180–187. doi: 10.1134/S0026261711020159

58. Ünal, B., Perry, V. R., Sheth, M., Gomez-Alvarez, V., Chin, K.-J., and Nüsslein, K. (2012). Trace elements affect methanogenic activity and diversity in enrichments from subsurface coal bed produced water. Front. Microbiol. 3:175. doi: 10.3389/fmicb.2012.00175

59. Veil, J. (2010). Water Management Technologies used by Marcellus Shale Gas Producers, ANL/EVS/R-10/3, Environmental

Environment Devision, Argonne National Laboratory for the U.S. Department of Energy, Office of Fossile Energy, National Energy Technology Laboratory. Chicago.

60.   Waldron, P. J., Petsch, S. T., Martini, A. M., and Nüsslein, K. (2007). Salinity constraints on subsurface archaeal diversity and methanogenesis in sedimentary rock rich in organic matter. Appl. Environ. Microbiol. 73, 4171–4179. doi: 10.1128/AEM.02810-06

61.   Walter, L. M., Budai, J. M., Martini, A. M., and Ku, T. C. W. (1997). Hydrogeochemistry of the Antrim Shale in the Michigan Basin. Des Plaines: Gas Research Institute, GRI-95/0251.

62.   Wang, Y., and Qian, P. Y. (2009). Conservative fragments in bacterial 16S rRNA genes and primer design for 16S ribosomal DNA amplicons in metagenomic studies. PLOS ONE 4:e7401. doi: 10.1371/journal.pone.0007401

63.   Whitman, W. B., Bowen, T. L., and Boone, D. R. (2006). "The methanogenic bacteria," in The Prokaryotes, Vol. 3, eds M. Dworkin, S. Falkow, E. Rosenberg, K.-H. Schleifer, and E. Stackebrandt (New York, NY: Springer), 165–207. doi: 10.1007/0-387-30743-5_9

64.   Wildgruber, G., Thomm, M., Konig, H., Ober, K., Ricchiuto, T., and Stetter, K. O. (1982). Methanoplanus-Limicola, a plate-shaped methanogen representing a novel family, the Methanoplanaceae. Arch. Microbiol. 132, 31–36. doi: 10.1007/BF00690813

65.   Wilharm, T., Zhilina, T. N., and Hummel, P. (1991). DNA-DNA hybridization of methylotrophic halophilic methanogenic bacteria and transfer of Methanococcus-Halophilus vp to the genus Methanohalophilus as Methanohalophilus-Halophiluscomb-nov. Int. J. Syst. Bacteriol. 41, 558–562. doi: 10.1099/00207713-41-4-558

66.   Wilhelm, E., Battino, R., and Wilcock, R. (1977). Low-pressure solubility of gases in liquid water. Chem. Rev. 77, 219–262. doi: 10.1021/cr60306a003

67.   Wuchter, C., Schouten, S., Coolen, M. J. L., and Sinninghe Damsté, J. S. (2004). Temperature-dependent variation in the distribution of tetraether membrane lipids of marine Crenarchaeota: implications for TEX 86 paleothermometry. Paleoceanography 19, PA4028. doi: 10.1029/2004PA001041

Chapter 3

Odour Removal In Leather Tannery

Rames C. Panda[1, 2], Chokalingam Lajpathi Rai[1],
Venkatasubramaniam Sivakumar[1], and Asit Baran
Mandal[1]

[1]Central Leather Research Institute (CSIR), Adyar, India
[2]Department of Chemical Engineering, Curtin University of Technology, Perth, Australia

ABSTRACT

Toxic odour causes pollution to environment. Removal of odour from wet processing sections of leather tanneries is important to preserve safety and occupational health. Such odour causing gases are identified in nature and are identified mostly as ammonia, Hydrogen sulfide and Volatile organic compounds. These gases, evolving from tanning drums, were experimentally quantified and analysed. Techniques for the abatement of odorous gases are designed using chemical methods. Scrubbing towers based on absorption of gases by liquid are designed and fabricated to evaluate the performance of the system in laboratory scale. Those gases were reduced in the concentration by absorbing through the packed bed vapour phase absorption using the activated carbon as the adsorbent. Results are helpful to conclude that the absorption technique presented here to reduce these toxic emission loads, seems to be simpler and economically cheaper.

INTRODUCTION

During the last two decades, cleaner production and pollution abatement have been developed which have made it possible to reduce the pollution load from tanneries. There are many techniques available (Riffenburg and Alison, 1941; IULTCS, 2008) [1,2] for treatment of wastewater. But less number of research are reported on pollution through air emission that give rises to bad odor and causes problem to occupational safety of the workers in Tannery. Hence the purpose of this work is to highlight sources of odor and to suggest techniques of their removal. In pre-tanning section of leather processing H_2S and NH_3 and VOC are evolved from process and cause bad odor. From putrefying hides also bad odor comes out. Other sources are from chemical storage in tannery (HCHO, CH_4 etc. gases comes out) and some organic foul gases form hides/skins during processing. In tannery, these gases either partly dissolved in process liquor, giving rise to odor or mix with air inside (tannery) and produces bad odor. In the latter case, biofilters, bioscrubbers, activated sludge scrubbers, trickling filter scrubbers, and reactive chemicals are available to suck and remove these bad odorous gases. But in the earlier case, suitable technology must be developed which is the objective of this paper. These emissions from tannery or effluent treatment plant, generally pollute the air, soil, surface water and underground water, causing serious health problems because they lie within and around residential areas. Respiratory disorders, diarrhea, dysentery and typhoid are the most serious illnesses among the community. The limiting steps for removal of ammonia—nitrogen and hydrogen sulphide are studied in production of wet blue hides.

SOURCES OF GASES

As mentioned earlier, the main constituents of odourous gases in tannery are VOC's (evolving during the action of enzymes causes decomposition and oxidation of the hides and skin), NH_3 and H_2S (evolving during the unhairing and deliming process of tanning in the processing of leather). Out of these the NH_3 and H_2S both are toxic gases which are the responsible for the odour in the tannery. The NH_3 having strong pungent and H_2S having fouling smell both giving a odour in and around the tannery environment. From Zahn et al. (2001) [3] it is found that a typical tannery will have VOC—100 ppm, NH_3—40 ppm and H_2S—30 ppm. Based on these loads, this work designs scrubbing systems for abatement of these toxic gases. According to ASTM standards the limits of these gases in air should be given by **Table 1**.

METHODS OF REMOVAL

Different methods of removal of NH_3 and H_2S are available: they are generally: addition of Chemical Reagents, Passing Compressed Air, Ozone Oxidation, Biochemical method.

Table 1: Air/water discharges, odor and exposure standards

Item	Concentration, mg/m³	
	H_2S	**NH_3**
(Air) Boundary—design ground level concentration	0.00014	0.6
(Air) Occupational health and safety time-weighted average	14	25
(Air) Short term exposure limit	21	35
Water	between 0.025 and 0.25 µg/l & sulphide = 1 mg/lit	< 0.78 mg/lit& kjeldahl nitrogen < 0.78 mg/li

Removal of Odor from Process Liquor

Different methods of removal of NH_3 and H_2S are studied: They are generally:

- Addition of Chemical Reagents;
- Passing Compressed Air;
- Ozone Oxidation;
- Passing air in counter current to liquor in a packed (activated carbon) bed;
- Biochemical and Biological methods.

Out of the above mentioned methods, the 4th one is economic and industrially feasible. Hence, trial is undertaken to study parametric effects on performance of removal of gases and design scale up.

Passing Compressed Air & Physical Absorption by Activated Charcoal

This method is developed at Bench Scale laboratory of Chemical Engg Dept, CLRI, Chennai, India, after many experimental trials. As packing material, charcoal bed is used. Samples of tannery wastewater are collected from different region and from different tanning sectors to analyze sulphide and nitrogen contents in them. Methods of removal of these odor-sources by means of adding reactive-chemicals or passing compressed air or adsorption by activated charcoal, are studied and discussed. Engineering design aspects of the proposed devices (CSTR/Packed Bed) are discussed. The odor removal process is made continuous by designing and employing a flow controller by adjusting flow rate of air passing through the float in a cstr/packed bed.

Experimental Setup

For the quantitative analysis the simpler and suitable setup was made. A column taken with the passage was taken. A fresh grade of Activated Carbon is being packed into that column up to the required height of the column. A pressure gauge meter is fixed across the height of the bed to calculate the pressure crop across the total bed height. The outlet of the column, i.e., exit of the air from column is made to collect in the sample collector for the detection and analysis. A two Round Bottomed flask were made mounted with the flow controller is attached to the bottom of column. From this the gas was produced artificially and passed through the bed. The experiment was conducted with analysis of two such gases, i.e., NH_3 and H_2S.

For Ammonia (NH₃)

The gas is artificially produced by means of heating up of the Ammonium hydroxide (NH_4OH) liquid at the desired concentration. It was made pass through the bed and collected out in the 3% of 50 cc of cold Boric acid solution. All the NH_3 gases passes out are caught up in the boric acid solution and the amount of ammonia in the collection sample is determined by quantitatively.

For Hydrogen sulfide (H₂S)

The gas is produced artificially by means of heating up of the Na_2S solution of desired concentration with the Hydrochloric acid (HCl) it produces the H_2S gas as

$$H_2S + 2NaOH \rightarrow NaHS + H_2O$$
$$NaOH + NaHS \rightarrow Na_2S + 2H_2O$$

The collected Na_2S product can be determined by quantitatively. In both the cases, the experiment was carried out in the different concentration and different temperature. The corresponding pressure drop at each case was recovered for the detection of respective quantity of gas. The time of exposure or passage through the bed was kept constant for all the cases to determine consistency at a temperature for all the cases.

Quantitative Analysis of the Gases

For NH_3

The collected NH_3 in the boric acid solution is shaken well colded up and then it is titrated against the N/10 HCl with phenolphthalein as indicator. From this the amount of concentration of NH_3 collected can be obtained.

For H_2S

The collected Na_2S sample is titrated against the N/10 $ZnSO_4$ solution. The titration is continued till all the Na_2S will converted to ZnS and Na_2SO_4 as

$$Na_2S + ZnSO_4 \rightarrow ZnS + Na_2SO_4$$

The end point is ascertained by spotting on to the dry filter paper previously impregnated with 10% lead acetate solution and the end point is reached when no immediate brown stain is visible due to conversion of all sulfides into zinc sulfate as the above equation.

In this case, removal efficiency is observed as 83%. A mathematical program was written to evaluate the process performance. The comprehensive procedure for analysis of design can be given as below:

Procedure

The steps for the process design are:

- Set on analysers (NH_3 and H_2S);
- Open air inlet valve (feed is air + NH_3 + H_2S mixed) that comes out from factory;
- Observe rotameter reading (about 20 psig). Air feeding rate is in gmol/min;
- Set desired air flow, water or liquid flow rates and connect analyzers to effluent gas stream. Record rotameter readings when it comes to steady state (generally it takes 5 mins);
- Read pressure drop across the column;
- Calculate NTU and HTU;
- Change the flow rates and observe the readings by repeating above steps.

MATHEMATICAL DEVELOPMENT

The schematic of gas absorption is shown in Figure 1. The odor gas flows from bottom to top of tower while liquid flows down from top to bottom of the packed column in counter-current.

Material Balance:

Gas Feed = (Air + NH$_3$ + H$_2$S) Liquid Feed = Water Counter current Gas Phase:

$$Lx_0 + Vy_F = Lx_B + Vy_1 \tag{1}$$

With x$_0$ = 0, we get

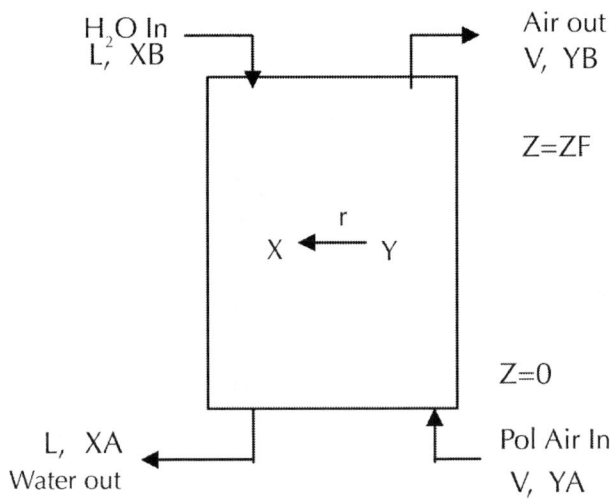

Figure 1: Schematic of gas absorption tower.

$$x_B = (V/L)(y_F - y_1) \tag{2}$$

Here V = vapor rate in g/min and L = Liquid rate in g/min.

We consider that the air (flow rate, V gmol/min) carrying the emitted gas stream enters the column of height z and diam D where it is scrubbed with water (flow rate = L gmol/min) in counter current. The rate of mass transfer is given by $r = k[y - y^*(x)]$ Where y and x are gas and liquid phase concentrations in mole fraction y*(x) is gas phase concentration equilibrium with liquid and can be evaluated by seeing VLE data. k is mass transfer coefficient. At steady state, gas phase and liquid phase transfers become equal and we get,

$$\frac{dy}{dz} = -\left(k/V\right)\left(y - y^*\right)$$

(3)

Integrating from $z = 0$ to $z = z_F$ we get

$$\frac{k}{V} z_F = \int_{y_B}^{y_A} \frac{dy}{y - y^*\left(x\right)}$$

(4)

= resistance to mass transfer. As the driving force $(y - y^*(x))$ becomes less and less, the integral becomes larger and larger, thus, indicating the Number of Transfer Units (NTU) becoming more and more. It is necessary to know $x(y)$ and $y^*(x(y))$ to evaluate the integral.

The Operating Line:

$$Lx_B + Vy = Lx + Vy_B \text{ or } x = \left(V/L\right)\left(y - y_B\right)$$

(5)

With $x_B = 0$, the integrand reduces to

$$\frac{1}{y - y^*\left(\left(V/L\right)\left(y - y_B\right)\right)}$$

(6)

which can be evaluated in terms of y if $y^*(x)$ is known. Generally, from VLE relationship we know,

$$y^* = K_H x$$

(7)

where K_H is called Henry's Law constant. Thus it is possible to find an analytical solution for the integrand and NTU and HTU can be calculated.

RESULTS AND DISCUSSION

The experimental data and readings are collected and analysed that resulted information given in Table 2. The molecules of the contaminated gas attracter to and accumulate on the surface of the activated carbon. Almost of the surface area available o the internal part structure and the adsorption depend on the pore volume and pore distribution. The amount of adsorption is totally dependent on the total surface area of the activated carbon.

The performance of removal of odourous components from gas streams by the column is shown in Figure 2. At 40°C, the tower can remove about 73% of NH_3 and 70.5% of H_2S from the odourous gas stream having a load of 800 ppm NH_3 and 200 ppm H_2S respectively. With increase in load the efficiency goes down.

Table 2: (a) Performance analysis from experimental studies of removal of odorous gases Time = 20 mins: Temperature = 40°C; (b) Performance analysis from experimental studies of removal of odorous gases Temperature = 40°C; (c) Performance analysis from experimental studies of removal of odorous gases Temperature = 45°C

(a)

Ammonia (ppm)	Pressure Drop	% Abatement
800	10.791	73.2
900	15.6	72.1
1000	19.6	71.3
Hydrogen Sulfide (ppm)	Pressure Drop	% Abatement
200	9.8	70.6
300	10.8	68.7
400	12.7	67

(b)

Ammonia (ppm)	Pressure Drop	% Abatement
800	25.5	72.40
900	19.6	70.8
1000	12.7	69.8
Hydrogen Sulfide (ppm)	Pressure Drop	% Abatement
200	10.8	69.5
300	12.5	68.1
400	14.7	66.7

(c)

Ammonia (ppm)	Pressure Drop	% Abatement
800	17.6	72.6
900	24.5	71.1
1000	30.4	69.2
Hydrogen Sulfide (ppm)	Pressure Drop	% Abatement
200	12.75	69
300	14.7	67.5
400	17.6	65.3

The removal efficiency can be increased to 90% by recycling and by intensify driving forces. It is predicted that by increasing number of horizontal partition/segments in the packed region, residence time of gases can be increased and thereby removal efficiency can be improved.

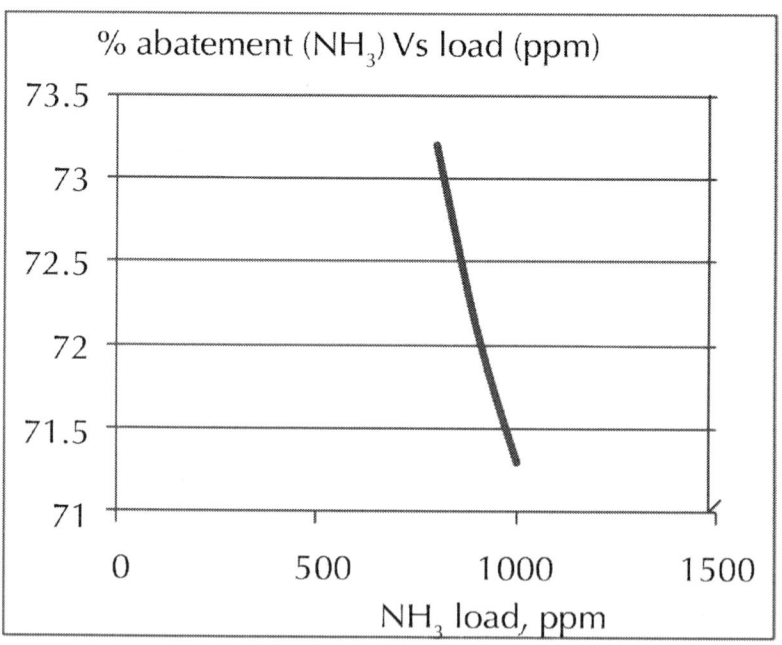

(a)

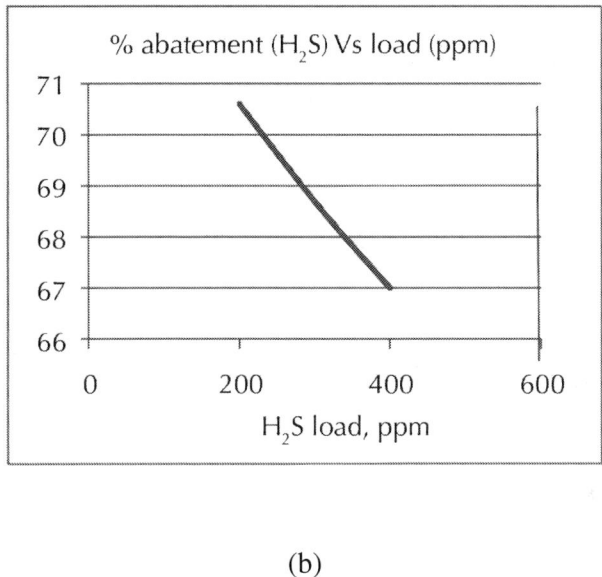

(b)

Figure 2: Performance of packed towers (removal efficiency of NH$_3$ and H$_2$S) at 40°C.

DESIGN OF COLUMN

The adsorption system is a simple one to design and it is made of a containment device, distribution and collection device to effect proper circulation of the gas stream through the activated carbon bed and a means for moving the gas stream through bed by using device like fan, blower or dispenser etc. The activated carbon bed can be conveniently configured into small drums or tanks depend on the size and application to the tannery. The superficial velocity Vs in ft/min is given by Vs = 48 (f − 1)/f where f = 1.2 (approx) and Actual gas flow rate, Q$_a$ can be calculated as: $Q_a = \dfrac{19.63Q(T+459)C_f}{P+14.2}$ where Q is gas folw rate in mmscfd, T is temperature in F, P is pressure in psig and C$_f$ is compressibility factor, 0.95. After finding out Q$_a$, the column size can be determined ID = (A/0.7854)$^{0.5}$ ft. The sizing for the application depends on the surface loading rate. The standard depth of minimum 4 ft carbon bed to maximum according to the gas flow rate and amount to be reduced. Larger flow of gas needs more height of bed in order to increase the surface area. The design includes the simple vessel or containment that can hold up the activated carbon.

Reactivation and Regeneration

Once the activated carbon is being used it has become saturated and therefore the adsorption decreases. In order to maintain the efficiency the activated carbon must be replaced with the fresh one after the every service of abatement. The carbon can be reactivated and regenerated by number of methods. The best method is washing the activated carbon in hot water and then dried off. It can be reused for the further process of adsorption.

The design calculations for the absorption tower is given as follows.

Calculations: Packed Tower Absorption Column

Data: Air flow rate = 7 cu ft/min

Temperature = 70 F

Pressure = 1 atm

Water flow rate = 500 ml/min

$x_B = 0$

NH_3 in air feed = 20,000 ppm

H_2S in air feed = 1000 ppm

NH_3 in air exit = 250 ppm

H_2S in air exit = 5 ppm

V = 7(cu ft/min)(28.32 L/ft3)(gmol/22.4 L)(492/530) = 8.125 gmol/min

L = 500(ml/min)(1 g/ml)(1 gmol/ 18 gm) = 27.8 gmol/min

$y_{A1} = (20,000/17)/(1,000,000/29) = 0.034$

$y_{A2} = (1000/34)/(1,000,000/29) = 0.0034$

$y_{B1} = (250/17)/(1,000,000/29) = 0.00043$

$y_{B2} = (05/34)/(1,000,000/29) = 0.0000172$

Equilibrium data: From Perry Handbook,

p_{NH_3} over 0.05 mole fraction aq NH_3 = 0.047

psia = 0.047/14.7 =0.0319 mole fraction NH_3 and 98.15/101.325 = 0.968 mole frac H_2S

For NH_3: K_H = y/x = 0.0319/0.050 = 0.638

For H_2S: K_H = y/x = 0.968/0.050 = 0.099 Operating Line:

$$x = (V/L)y - y_B = 0.296y - 0.00043$$

$$\text{and} \quad y - y^* = y - K_H x$$

$$= y - 0.638(0.296y - 0.00043)$$

$$= 0.818y + 0.000274$$

$$NTU = \int_{0.00042}^{0.034} \frac{dy}{y - y^*(x)} = 4.70223$$

(8)

It also looks into the possibility of the usage of controllers for effective treatment process. With variation of process recipe and quality of hides, sulphide content of process liquor may vary at the exit. Hence to balance the load changes, an air flow controller is necessary. Accordingly, air flow rate is controlled/adjusted to maintain the emission standards.

CONCLUSIONS

Chemical engineering methods are used to reduce odourous gas loading from leather industry. From the above study it was clear that the adsorption of the toxic gases such as ammonia and H_2S are quantitatively reduced by passing through vapour phase adsorption. Here it was concluded that the application of the vapour phase adsorber with activated carbon would become the simple and economically cheaper in all the aspects from the installation to the running. And also it was clear that this method would give the better results in the reduction of odour in the effluent. This could be a better method if implemented in the end process of the effluent treatment to reduce the concentration of the contaminants and thus helpful for reducing pollution and global warming.

ACKNOWLEDGMENTS

The authors are willing to acknowledge the help from network project (NWP-044) to carry-out the experiments.

REFERENCES

1.   IULTCS, "IUE-8: Recommendations for Odor Control in Tanneries," 2008. http://www.iultcs.org/pdf/IUE8_2008.pdf

2.   H. B. Riffenburg and W. W. Alison, "Treatment of Tannery Wastes with Flue Gas and Lime," Industrial & Engineering Chemistry Research, Vol. 33, No. 6, 1941, pp. 801-803.doi:10.1021/ie50378a026

3.   J. A. Zahn, et al., "Correlation of Human Olfactory Responses to Airborne Concentrations of Malodorous Volatile Organic Compounds Emitted from Swine Effluent," Journal of Environmental Quality, Vol. 30, No. 2, 2000, pp. 624-634. doi:10.2134/jeq2001.302624x

Effect of Two Liquid Phases on the Separation Efficiency of Distillation Columns

Gardênia Marinho Cordeiro, Stephanie Rolim Dantas, Luís Gonzaga Sales Vasconcelos, and Romildo Pereira Brito

Department of Chemical Engineering, Federal University of Campina Grande, Campina Grande, Brazil

ABSTRACT

Distillation is one of the oldest and most important separation processes used in the chemical and petrochemical industries. On the other hand, it is a process the thermodynamic efficiency of which is very low, and therefore reducing the consumption of energy is one of the targets of research studies on distillation. This article arose from seeking to reduce energy consumption in a distillation train of 1,2-dichloroethane (ethylene dichloride-EDC) of a commercial plant producing vinyl monochloride (VMC), which involves an azeotropic distillation column. The reduction in the reboiler heat duty caused significant changes in concentration and temperature profiles throughout the column due to the formation of two liquid phases. The results show that, although very small in percentage terms (less than 2.5%), the appearance of the 2nd liquid phase causes significant changes in the operation of the column and the separation achieved.

INTRODUCTION

Distillation is one of the oldest and most important separation processes used in chemical processes. On the other hand, its thermodynamic efficiency is extremely low, which accounts for the high percentage of global energy consumed in a plant. In general, distillation column reboilers consume over 50% of the energy involved in the process of heat exchange in a plant (Soave and Feliu, 2002 [1]).

The term azeotropic distillation is applied to the class of techniques based on fractional distillation in which azeotropic behavior is exploited to achieve separation. Traditionally, the specie that causes the azeotropic behavior is added as a mass separating agent: the entrainer. In some situations it may be present in the feed mixture (self-entraining) of the azeotropic column (Perry et al., 1999 [2]).

Although a large number of studies involve azeotropic distillation, most involve columns in which a third component is added in order to further the separation. Such studies are about choosing the third component, the influence of a thermodynamic model, evaluating the existence of multiple steady states and the study of process control (Laroche et al., 1992 [3]; Bekiaris et al., 2000 [4]; Magnussen et al., 1979 [5]; Rovaglio and Doherty, 1990 [6]; Wang et al., 1997 [7]; Luyben, 2008 [8]; Wu and Chien, 2009 [9]). Another striking feature of the articles cited is that they consider the formation of two liquid phases only in the reflux vessel.

Lao and Taylor (1994) [10] reviewed the literature on the separation efficiency of distillation columns involving three-phase systems, and cite several sources which give rise to their finding that the conclusions drawn on these systems are contradictory. Some studies claim that overall efficiency was not influenced by the number of liquid phases present. Other studies indicate that the introduction of a second liquid phase may have a strong (positive or negative) influence on the mass transfer.

Widagdo and Seider (1996) [11] published one of the most complete (and even to this day, one of the most cited) articles on the azeotropic distillation process. They showed that knowledge contained in the literature is scant both as to a real understanding of the process and the difficulties regarding control of azeotropic columns. They also emphasized the issue of the formation of two liquid phases within the column, but there is no consensus on the efficiency of separation when columns operating with one and with two liquid phases are compared.

In 1997 Wang et al. [7] observed experimentally the formation of two liquid phases inside a column, depending on the reflux and the reboiler heat duty, as well as the presence of multiple steady states; the study evaluated the dehydration of isopropanol, using cyclohexane as the entrainer.

According to Higler et al. (2004) [12], azeotropic distillation is characterized by its operational complexity, due to the possible formation of two liquid phases inside the column. The authors used an equilibrium and a non-equilibrium stage model and claimed the formation of two liquid phases in the distillation column influences the mass transfer process, thus affecting efficiency.

The equilibrium stage model, widely used in modeling and simulating distillation processes, does not represent the reality that few stages actually operate in equilibrium. This problem can be solved by introducing Murphree efficiencies. However, some authors (Cairns and Furzer, 1990 [13]) warned against incorporating Murphree efficiencies into equilibrium stage models of three-phase systems. In fact, the projections may be more accurate if a non-equilibrium stage model is considered. However, calculations are complex, thus requiring more computational time, which is not desirable for control applications. But, the biggest obstacle is that the parameters required to perform the calculations are rarely available.

Junqueira et al. (2009) [14] analyzed the formation of two liquid phases in the azeotropic column in the production of anhydrous ethanol, and, in order to decrease this phenomenon, many process configurations have been studied as well as variations in operating conditions. It was concluded that the formation of the second liquid phase may affect the performance of the column and consequently reduce its efficiency.

Silva et al. (2003) [15] evaluated the dynamics of an azeotropic distillation column similar to the one considered in this article; however, the entrainer was already present in the feed, which was held in the intermediate region of the column, and the formation of two liquid phase occurred only in the reflux vessel.

Guedes et al. (2007) [16] followed the same procedure as the one studied in this paper and, in the steady state, evaluated the process sensitivity relative to the feed temperature; and, dynamically evaluated the influence in feed temperature, including the operation condition with two liquid phases in some plates.

The distillation column considered in this article shows characteristics of an azeotropic distillation, since two liquid phases form in the reflux vessel and, depending on the operation condition, in some stages throughout the column. However, another feature makes the system unconventional: the feed takes place in the reflux vessel. In the research literature few studies have considered systems with these characteristics.

PROBLEM STATEMENT

The distillation column considered in this study is part of the purification train of 1,2-dichloroethane (ethylene dichloride-EDC) of a commercial plant which produces vinyl chloride monomer (VCM).

The process of obtaining EDC occurs through the direct chlorination of ethylene (C_2H_4), as shown in the reaction: $C_2H_4 + Cl_2 \rightarrow C_2H_4Cl_2$. The EDC product (high purity) leaves the reactor and moves on to the purification system, where it undergoes aqueous washing. Figure 1 shows the flow diagram of the EDC dehydration process, where it can be observed that aqueous washing is conducted in the separating vessel (or reflux vessel). After the top condenser and in the reflux vessel, there are two liquid phases: an organic one, saturated in H_2O, and an aqueous one, saturated in organic matter. The organic phase returns to the reflux of the column, while the stream of the aqueous phase is discarded.

Although less volatile than the EDC, the H_2O leaves from the top of the column due to the reversal in the value of the constant K (Figure 2), which is due to the fact that H_2O forms a minimal azeotrope, not only with the EDC, but with almost all organic compounds present in the process.

Note that in the stream coming from the reactor (FROMR1) there is no H_2O, so that during washing, the stream that carries out the reflux of the column (TODRY 2) becomes saturated in H_2O.

A close analysis of Figure 1 leads to the conclusion that the system as a whole can be seen as a conventional column (with reboiler, condenser and reflux vessel), with the feed (FROMR1 and WATER) in the reflux vessel. In industry, although the analysis of the degree of freedom indicates two variables will be manipulated, only the reboiler heat duty is used, since the reflux flow rate is used to control the level (organic phase) of the vessel and the distillate flow rate (WASTE) cannot return to the process.

The study by Guedes et al. (2007) [16] aimed at reducing the consumption of energy in the azeotropic column. The question to be answered was: if the reboiler heat duty is the only manipulated variable used, to what extent can it be reduced without compromising the quality of the bottom product (the H_2O mass fraction)?

Accordingly, by performing tests in the plant, the reboiler heat duty was gradually reduced, which resulted in plate temperatures (top, middle and bottom) that were much smaller than those observed historically, being indicated. In spite of the amount of moisture in the bottom stream being below the specification (10 ppm), the tests were discontinued after 7 hours of operation, and a new operating condition (lower heat duty) was established.

According to Guedes et al. (2007) [16], a more significant change in the temperature profile occurs because of the formation of a 2nd liquid (aqueous) phase in the plates of the column. And, the good agreement between azeotropic data (Azeotropic Data, 1973 [17]) and solubility (Dechema, 1990) found in the literature for EDC-H_2O and those predicted by the simulations, are the mainstays of this conclusion. However, the simulations were carried out without formally defining an objective function and constraints (optimization). Furthermore, no evaluation of the effect of the possible presence of a 2nd liquid phase in separation was performed. Thus, this study aimed to: formalize optimizing the consumption of energy, and evaluate the efficiency of separation taking two operating conditions into account: without the formation of two liquid phases (Case I) and with the formation of two liquid phases (Case II).

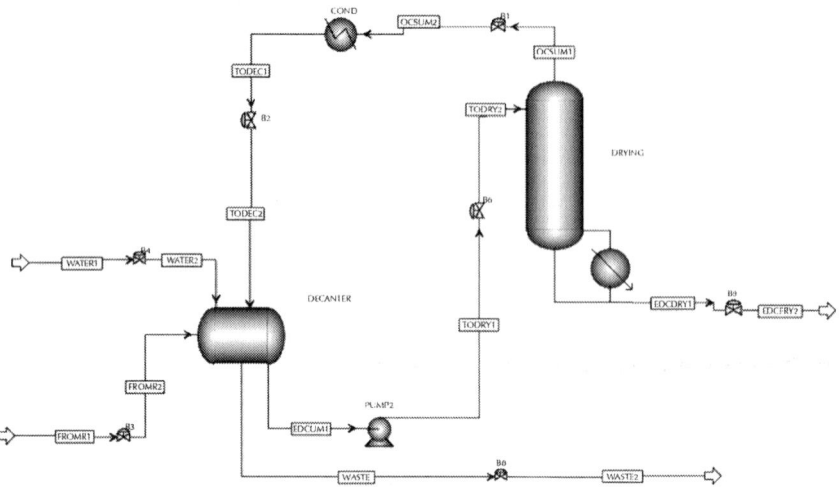

Figure 1: Flowsheet of the EDC dehydration process.

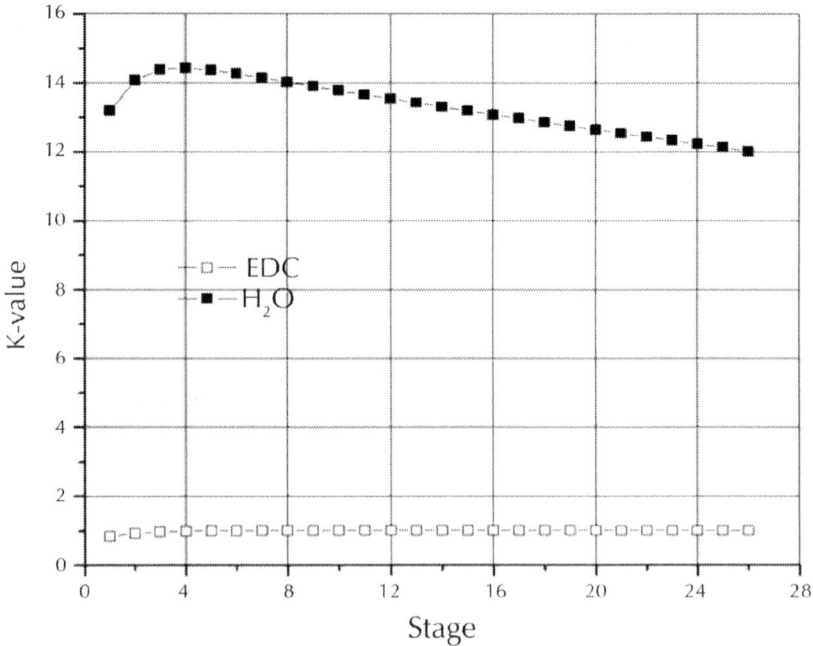

Figure 2: K-values along a column working with a single liquid phase.

MODELING AND SIMULATION

The simulation was performed using Aspen Plus™ commercial simulator. In order to represent the real system, the system was modeled using reboiled absorption, followed by a condenser (Heater) and a decanter (Decanter). To model the column in question, the RadFrac™ routine was used.

The RadFrac™ routine detects the possible formation of a second liquid phase (the main component was H_2O) at any stage; assumes there is an equilibrium stage model; and uses specified values for stage efficiencies. These efficiencies can be manipulated to adapt to the plant data. In this study, a Murphree efficiency equal to 64% for all plates and 100% for the reboiler was used. In the industrial plant, the column has 25 stages (numbered from top to bottom) and a reboiler type thermosyphon. In the Aspen Plus™ simulator, the pressure in each plate of the column, as well as in the other equipment, is kept constant.

To represent the equilibrium between liquid-liquidvapor phases (ELLV), a - procedure was used. Even with the column operating under low pressure, the vapor phase was represented by the Redlich-Kwong Equation of State (EOS). The activity coefficient was determined from the NRTL model (Perry et al., 1999 [2]), which represents the ELLV system effectively. Tables 1 and 2, respectively, show the comparison between the azeotropic (Azeotropic Data, 1973) and solubility data (Dechema, 1990 [18]) found in the literature for the EDC-H_2O system (main components) and those predicted by the simulations.

In order to determine the optimal energy consumption, the objective function (J) to be minimized was defined as the reboiler heat duty (Qr).

The restriction in the case of optimization without the presence of two liquid phases (Case I) is the mass fraction of H_2O in the liquid phase (global): if it was not desired to form two liquid phases over the column, the restriction imposed was 2500 Parts Per Million (ppm) (approximately the saturation value of EDC with H_2O at 45˚C) for the first stage (numbered from top to bottom) of the column. The choice of this plate was due to its being found that the formation of two liquid phases starts in this plate.

For the operation with two liquid phases (Case II), the restriction imposed was 10 ppm in the bottom stream of the column (the maximum permitted in the plant). Mathematically, the problem was formulated as follows:

$$\text{Min } J = Qr$$

(1)

Subject to

$$x_{\#1}^{H_2O} \leq 0.0025$$

(2)

Or

$$x_{Bott}^{H_2O} \leq 0.00001$$

(3)

The optimization procedure considered the distillate flowrate (stream OCSUM1) as the manipulated variable (OCSUM1). The objective function was inserted via the Analysis/Optimization Model of the Aspen Plus™ tool, which uses the Sequential Quadratic Programming (SQP) search method for the optimum. The restrictions were inserted using the Analysis/Constraint Model.

The procedure can be implemented over the following steps:

- Fix the number of stages of the column;
- Specify the value of the distillate flowrate, which will be used as an initial estimate;
- Insert, via the Analysis/Optimization Model, the objective function and the range over which the variable may be manipulated;
- Insert, via the Analysis Constraint Model, the restriction and its tolerance.

Table 1: Comparison of azeotropic data for EDC (1)-H_2O (2) system

Azeotropic boiling Point (1 atm), °C		Mass Traction of H_2O	
Literature	Aspen Plus™	Literature	Aspen Plus™
71.85	73.85	9.2	9.6

Table 2: Solubility (% weight) of EDC (1)-H_2O (2) system

Temparature,°C	Literature		Aspen Plus™	
	(1) in (2)	(2) in (1)	(1) in (2)	(2) in (1)
30	0.889	0.163	0.888	0.163
40	0.948	0.213	0.940	0.210
50	1.040	0.286	1.023	0.279
60	1.170	0.391	1.149	0.379
70	1.337	0.529	1.331	0.526

STEADY-STATE RESULTS

A comparison of data from the plant (the historical operating conditions) and those provided by the simulation is shown in Table 3. The good agreement between real and simulated data, in fact, proves the effectiveness of the modeling and the simulation.

Table 4 shows the conditions of the stream from the reactor (FROMR1) and Table 5 presents results for two operating conditions: 1) historical and 2) optimized.

As per Table 5, with the formation of two liquid phases (Case II), the reduction in energy consumption compared with the situation with a single liquid phase (Case I) is 19.4%; a result caused by a decrease in the distillate flow rate.

The final value of the reboiler heat duty was derived and determined after the constraints were optimized. In both cases, the production of "dry" EDC (EDCDRY2) was very similar.

In Figure 3, note the large difference between the temperature profiles for the two optimal situations. For Case I, a significant variation occurs between the 1st and the 5th plate, and then the rate of increase is almost linear from there to the 26th plate (bottom). On the other hand, in case II, the variation in the rate of increase between the 1st and 16th plate is almost linear, and then there are steep increases in this rate until the 24th plate at which point the temperatures in the two cases coincide.

Table 3: Comparison between the real and simulation data (Guedes et al., 2007)

Variable	Real	Simulation
Reboiler heat duty(Kcal/h)	1.52×10°	1.53×10°
Temparature at top (ºC)	79.0	79.4
Temparature of plans 6(ºC)	85.0	87.0
Temparature at bottom (ºC)	93.0	93.4

Table 4: Characteristics of the feed (FROMR1)

	Value
Temparature ºC	40.0
Flowrate, Kg/h	59,250
Mass fraction	
1,1-dichloroethane	0.00009
Carbon-tetracloride	0.00002
1,2- dichloroethane(EDC)	0.99398
Water	0.00000
1,1,2-trichloroethane	0.00130
1,1,3-trichloroethane	0.00461

Table 5: Results for two operational conditions

Historical		Optimized	
Case I		Case II	
Distillate flowrate(kg/h)	4850.0	4616.9	1465.3
Reboiler heat duty(kcal/h)	1.52×10°	1.4985×10°	1.2079×10°

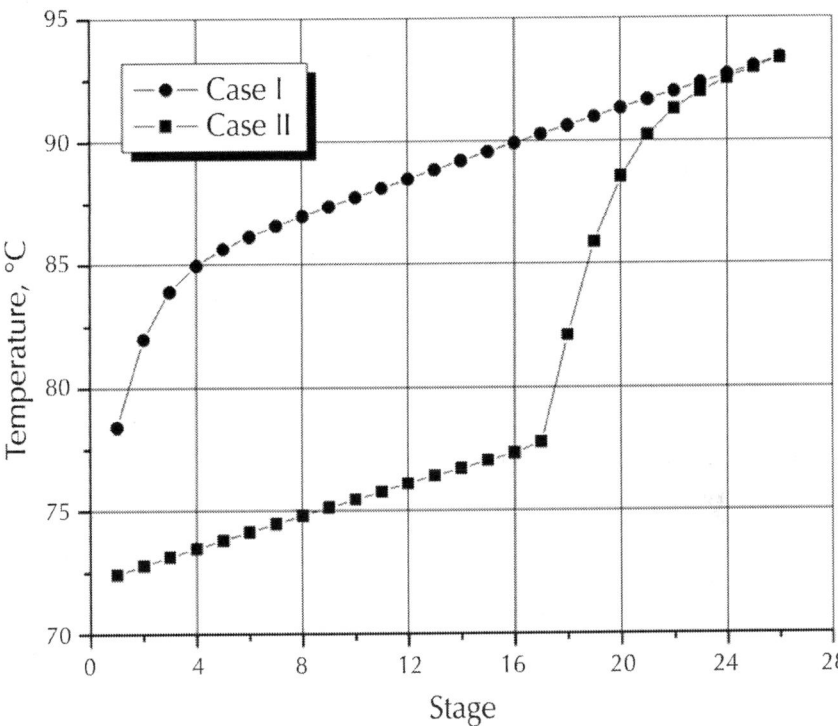

Figure 3: Temperature profiles for the two optimized situations.

In both cases, the linear behavior of the temperature takes place basically by varying the pressure, since the change in the composition of the species along the column is very small, as shown in Figure 4. Simulations that include a negligible pressure drop along the column show the temperature profiles then remain on plateaus, rather than go straight upward, thus confirming this observation on the result of there being negligible drops in the pressure. The profiles obtained experimentally by Wang et al. (1997) [7] show qualitative forms similar to Figure 3. However, unlike the findings of this study, the percentage of H_2O present in the feed was high.

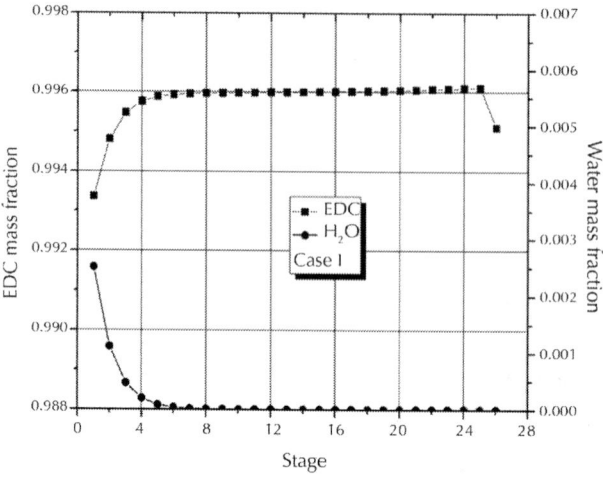

(a)

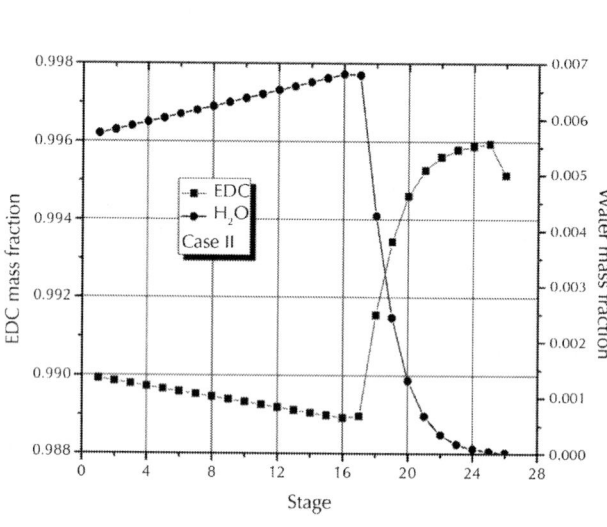

(b)

Figure 4: (a) Composition profile (EDC and H$_2$O) in the liquid phase (global) for Case I; (b) Composition profile (EDC and H$_2$O) in the liquid phase (global) for Case II.

Figure 4 shows the mass fraction of EDC and H_2O (main components) in each stage, from which it may be seen that, in each case, the mass transfer is at its most significant in different regions of the column: for Case I in the upper region; for Case II, in the lower one. For Case II, the greatest change in composition occurs in the region where the 2nd liquid phase is not present (from the 16th stage on). In fact, in both cases, dehydration mainly occurs in a small region of the column.

Given the low transfer of mass in most of the column, Figure 4 suggests that the number of stages of the column could be smaller. In fact, if the reboiler heat duty is maintained constant, simulations for a column with 19 stages show the presence of a single liquid phase and the fraction of H_2O at the bottom is within specification. However, for columns with 18 stages, two phases are present and the liquid fraction of H_2O at the bottom (1000 ppm) is above the one laid down in the specification.

The reason for the formation of two liquid phases can be seen in Figure 4. For Case II, in the region of two liquid phases, the maximum mass fraction of H_2O is about 0.7% by weight, so it is above the saturation value of the organic phase with H_2O. For Case I, the maximum mass fraction of H_2O is around 0.25% by weight (approximately the saturation value of EDC with H_2O). The behavior of Case II is due to the fact that the decrease in the reboiler heat duty does not prompt the removal of H_2O (in the form of azeotrope) in the early stages of the column.

Figure 5 shows the Separation Factor (SF) defined by Equation (4) Perry et al., 1999 [2]) along the column, in which what can be noted is that the separation efficiency is increased when there is a single liquid phase. Even if the second liquid phase is present, the Separation Factor is greater in stages where this phase disappears. From this Figure, note also that, for Case II (a two liquid phase up to plate 16), dehydration occurs in the last few plates. Overall, the magnitude of the Separation Factor measured for Case I (1.15E9) was completely different from that calculated for Case II (235).

$$\frac{y_{H_2O}/x_{H_2O}}{y_{EDC}/x_{EDC}} \tag{4}$$

The reduction in the SF for Case II may be explained as a direct consequence of the reduction of the reflux flow rate (caused by the decreased flow of distillate), which is usually one of the variables that most impact separations. However, what needs to be taken into account is that a simulation condition which operates immediately before the 2nd liquid phase forms and which involves a minimal reduction in the reflux flow rate, results in an SF of 1.7E9, that is, in the same order of magnitude of that calculated for Case I.

This result is in accordance with various citations in the article by Widagdo and Seider (1996) [11] and as pointed out by Junqueira et al. (2009) [15]. That is to say there is a drastic reduction in the separation efficiency of columns operating with two liquid phases in some plates.

The results presented in Figure 5 were obtained after optimizing the reboiler heat duty and assuming a constant Murphree efficiency (64%). Figure 6 shows the global Separation Factor H_2O/EDC depending on the Murphree efficiency, without considering the optimization. For operation with a single liquid phase (Case I) a distillate flow rate was set at 3 500 kg/h, while the condition for the two liquid phases (Case II), this was set at 1 450 kg/h.

From the results of Figure 6, it is possible to note that the operation with a single liquid phase in the plates is much more dependent on the operational efficiency in which some of the column plates have two liquid phases.

For Case I, the behavior is similar to that typically observed for distillation columns: separation is directly proportional to the efficiency of the stages. Moreover, where two liquid phases are observed in some plates of the column (Case II), separation decreases when the efficiency of the plate is increased, which is caused by increasing the number of plates with two liquid phases (16 to 18).

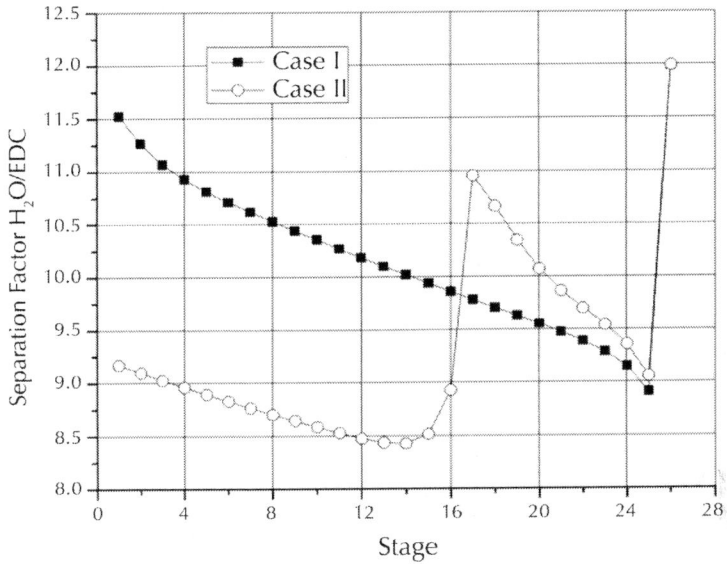

Figure 5: Separation factor H_2O/EDC along the column for two optimized situations.

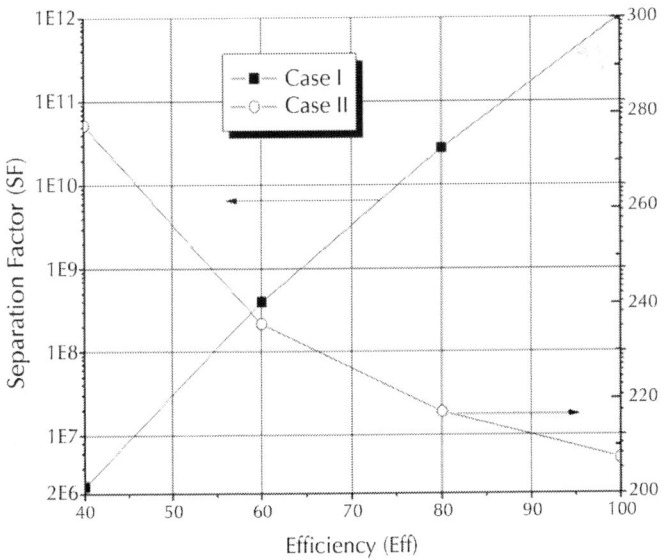

Figure 6: Global separation factor H_2O/EDC for two optimized situations.

CONCLUSIONS

Using as a case study the dehydration of 1, 2-dichloroethane (EDC) of a commercial plant that produces vinyl chloride monomer—VCM, the study aimed to evaluate the separation efficiency for two operating conditions: one with two liquid phases (Case II) and one with a single liquid phase (Case I) throughout the stages of an azeotropic distillation column.

Although very small as a percentage (less than 2.5%), the appearance of the 2nd liquid phase causes significant changes in the operation of the column and the separation achieved.

In each case, the mass transfer is at its most significant in different regions of the column: for Case I, in the upper region, for Case II, in the lower one. In fact, the transfer of mass increases when the 2nd liquid phase is not present, that is, the separation efficiency is increased when there is a single liquid phase present.

It is not a reduction in the reflux that causes the strong decrease in the Separation Factor (Case I compared to Case II); in fact, the drastic reduction in the efficiency of separation is the result of the operation with two liquid phases in some plates of the column.

ACKNOWLEDGEMENTS

The authors are grateful to the Brazilian National Council for Scientific and Technological Development (CNPq) for their financial support, and also to Braskem for permission to publish the results of this study.

REFERENCES

1.   G. Soave and J. A. Feliu, "Saving Energy in Distillation by Feed Splitting," Applied Thermal Engineering, Vol. 22, No. 8, 2002, pp. 889-896.

2.   R. H. Perry, D. W. Green and J. O. Maloney, "Perry's Chemical Engineer's Handbook," 7th Edition, McGrawHill, New York, 1999.

3. L. Laroche, N. Bekiaris, H. W. Andersen and M. Morari, "The Curious Behaviour of Homogeneous Azeotropic Distillation— Implications for Entrainer Selection," AIChE Journal, Vol. 38, No. 9, 1992, pp. 1309-1328.

4. N. Bekiaris, E. G. Guttinger and M. Morari, "Multiple Steady States in Distillation: Effect of VL(L)E Inaccuracies," AIChE Journal, Vol. 46, No. 5, 2000, pp. 955-979.

5. T. M. Magnussen, L. Michelsen and A. A. Fredenslund, "Azeotropic Distillation Using UNIFAC," Chemical Engineering Progress Symposium Series, Vol. 56, No. 4, 1979.

6. M. Rovaglio and F. M. Doherty, "Dynamics of Heterogeneous Azeotropic Distillation Columns," AIChe Journal, Vol. 36, No. 1, 1990, pp. 39-52.

7. C. J. Wang, D. S. Wong, I.-L. Chien, R. F. Shih, S. J. Wang and C. S. Tsai, "Experimental Investigation of Multiple Steady States and Parametric Sensitivity in Azeotropic Distillation," Computer and Chemical Engineering, Vol. 21, 1997, pp. S535-S540.

8. W. L. Luyben, "Control of the Heterogeneous Azeotropic n-Butanol/Water," Energy and Fuels, Vol. 22, No. 6, 2008, pp. 4249-4258. doi:10.1021/ef8004064

9. Y. Wu and I. Chien, "Design and Control of Heterogeneous Azeotropic Column System for the Separation of Pyridine and Water," Industrial & Engineering Chemistry Research, Vol. 48, No. 23, 2009, pp. 10564-10576. doi:10.1021/ie901231s

10. M. Z. Lao and R. Taylor, "Modeling Mass-Transfer in 3- Phase Distillation," Industrial and Engineering Chemistry Research, Vol. 33, No. 11, 1994, pp. 2637-2650.doi:10.1021/ie00035a015

11. S. Widagdo and W. D. Seider, "Azeotropic Distillation," AIChE Journal, Vol. 42, No. 1, 1996, pp. 96-130.

12. A. Higler, R. Chande, R. Taylor, R. Baur and R. Krishna, "Non-Equilibrium Modeling of Three-Phase Distillation," Computers and Chemical Engineering, Vol. 28, No. 10, 2004, pp. 2021-2036. doi:10.1016/j.compchemeng.2004.04.008

13. B. P. Cairns and I. A. Furzer, "Multicomponent 3-Phase Azeotropic Distillation—Extensive Experimental Data and Simulation Results," Industrial and Engineering Chemistry Research, Vol. 29, No. 7, 1990, pp. 1349-1363. doi:10.1021/ie00103a040

14. T. L. Junqueira, M. O. S. Dias, R. Maciel Filho, M. R. W. Maciel and C. E. V. Rossel, "Simulation of the Azeotropic Distillation for Anhydrous Bioethanol Production: Study on the Formation of a Second Liquid Phase," Computer Aided Chemical Engineering, Vol. 27, 2009, pp. 1143-1148. doi:10.1016/S1570-7946(09)70411-0

15. A. R. Silva, J. H. P. Brooman, L. R. Braga Jr., L. G. S. Vasconcelos and R. P. Brito, "Steady-State and Dynamics Behavior of an Industrial Azeotropic Distillation Column," The 6th Italian Conference on Chemical and Process Engineering, Pisa, 8-11 June 2003.

16. B. P. Guedes, M. F. Figueiredo, L. G. S. Vasconcelos, A. C. B. Araújo and R. P. Brito, "Sensitivity and Dynamic Behavior Analysis of an Industrial Azeotropic Distillation Column," Separation and Purification Technology, Vol. 56, No. 3, 2007, pp. 270-277. doi:10.1016/j.seppur.2007.02.014

17. "Azeotropic Data-III, Advances in Chemistry Series," In: R. F. Gould, Ed., Advances in Chemistry Series, Vol. 116, American Chemical Society, Washington DC, 1973, pp. 1-6.

18. Pennsylvania State University, "Dechema Chemistry Data Series," Deutsche Gesellschaft für Chemisches Aparatewesen, Frankfurt am Main, 1990.

Chapter 5

Removal of Cesium on Polyaniline Titanotungstate as Composite Ion Exchanger

I. M. El-Naggar[1], E. S. Zakaria[1], I. M. Ali[1], Magdy Khalil[1], and M. F. El-Shahat[2]

[1]Atomic Energy Authority, Hot Labs. Center, Cairo, Egypt
[2]Chemistry Department, Faculty of Science, Ain Shams University, Cairo, Egypt

ABSTRACT

Polyaniline titanotungstate (PATiW) was synthesized by the sol-gel method. Adsorption isotherm studies of Cs^+ from aqueous solution are described. Elemental Composition, chemical solubility, ion-exchange capacity (IEC) and pH titration are studied. Distribution coefficients (K_d) for ten metal ions have been determined. It was found that the polyaniline titanotungstate has high affinity and high selectivity for Cs^+. The material was high separation of Cs^+ from other metal ions. The adsorbent capacity was determined using the Freundlich and Langmuir adsorption isotherm models. The Cs^+ adsorption isotherm data fit best to the Freundlich isotherm model. The maximum Cs+ uptake of polyaniline titanotungstate was found 217 mg/g. A column tests were performed to determine the breakthrough curves with varying bed depths and flow rates in different solutions. The results show that the half breakthrough time increases proportionally with increasing bed depths. Kinetic studies for removal cesium from milk were investigated.

INTRODUCTION

Radioactive waste is an inevitable residue from the use of radioactive materials in industry, research and medicine, as well as from the use of nuclear power to generate electricity. The management and disposal of such waste is, therefore, an issue relevant to almost all countries. The ever increasing pressure to reduce the release of radioactive and other toxic substances into the environment requires constant improvement/ upgrading of processes and technologies for treatment and conditioning of liquid radioactive waste. Treatment of liquid radioactive waste quite often involves the application of several steps such as filtration, precipitation, sorption, ion exchange, evaporation and/or membrane separation to meet the requirements both for the release of decontaminated effluents into the environment and the conditioning of waste concentrates for disposal. New materials and processes are under consideration and development in various countries [1]. Most of the combined radioactivity in liquid nuclearwaste is a result of the fission products ^{137}Cs and ^{90}Sr. ^{137}Cs is an important radiocontaminant with its long half life ($t_{1/2}$ = 30.17 years) and represents a serious radiological hazard because as an alkaline element, it is easily assimilable by living organisms [2]. The ion exchange is the most important method for the selective adsorption and safe storage of ^{137}Cs. Although a number of organic exchangers exist, which are selective towards cesium, they are easily decomposed when exposed to highly ionizing radiation. Inorganic ion exchangers have several superior qualities required for the treatment of nuclear waste effluents compared to organic resins. In order to obtain a combination of these advantages associated with polymeric and inorganic materials as ionexchangers, attempts have been made to develop polymeric-inorganic composite ion-exchangers by incorporation of organic monomers in the inorganic matrix [3]. Few such excellent ion-exchange materials have been developed in our laboratory and successfully being used in chromatographic techniques [4-6]. An inorganic precipitate ion-exchanger based on organic polymeric matrix must be an interesting material, as it should possess the mechanical stability due to the presence of organic polymeric species and the basic characteristics of an inorganic ion-exchanger regarding its selectivity for some particular metal ions [7-10]. It was therefore considered to synthesize such hybrid ion-exchangers with a good ion-exchange capacity, high stability, reproducibility and selectivity for metal ions, indicating its useful environmental application.

Efforts have been made to improve the chemical, thermal and mechanical stabilities of ion exchangers and to make them highly selective for certain metal ions. Silica potassium cobalt hexacyanoferrate composite ion exchanger has excellent exchange properties of cesium [11]. Polyaniline Ce(IV) molybdate has high selectivity to Cd(II) [12]. Poly-o-methoxyaniline Zr(IV) molybdate has high selectivity to Cd(II) [13]. Acrylonitrile stannic(IV) tungstate has high selectivity to Pb(II) [14]. Cobalt ferrocyanide impregnated organic anion exchanger was found to be highly selective for cesium [15].

In the following parts of this contribution, composite absorbers, titanotungstate with polyaniline binding matrix was prepared and its properties and technological application are evaluated. Ion exchange capacity and distribution coefficient were determined towards cesium. Isotherm and column studies were applied at different conditions for removal cesium. It is found high selectivity for cesium from other metal ions from radioactive waste, aqueous solution and applied for removal cesium from milk.

EXPERIMENTAL

Chemicals and Reagents

All chemicals and reagents used in this work were of analytical grade purity and used without further purification. Cesium chloride, titanium tetrachloride, $TiCl_4$. H_2O, potassium persulphate ($K_2S_2O_8$) were obtained from Prolabo (England). Sodium tungstate $Na_2WO_2 \times 2H_2O$, aniline ($C_6H_5NH_2$), nitric acid and hydrochloric acid were purchased from Adwic (Egypt).

All samples and chemicals used in this work were weighted using an analytical balance of Bosch type having maximum sensitivity of 10^5 g and accuracy ±0.001/y. For the equilibrium experiments, a good mixing for the two phases was achieved using thermostatic shaker water bath of the type Julabo SW-20C obtained from West Germany.

Preparation of Polyaniline Titanotungstate (PATiW)

Polyaniline gels were prepared by mixing aqua volumes of the solutions of 10% aniline ($C_6H_5NH_2$) and 0.1 M potassium persulphate with continuous stirring by a magnetic stirrer. Green colored polyaniline gels were obtained by keeping the solutions below 10°C for half an hour [16]. A precipitate of titanium tungstate was prepared at (65°C ± 1°C) by adding 1 M titanium chloride solution to an aqueous solution of 1 M sodium tungstate ($Na_2WO_2 \times 2H_2O$) in equal volume ratio. The white precipitates were obtained, when the pH of the mixtures was adjusted to 6.5 by adding aqueous ammonia with constant stirring. The gels of polyaniline were added to the white inorganic precipitate of titanium and mixed thoroughly with constant stirring. The resultant green colored gels were kept for 24 h at room temperature (25°C ± 1°C) for digestion. The supernatant liquid was decanted and the gel was rewashed with bidistilled water in order to remove fine adherent particles and was filtered by center fission. The excess acid was removed by washing with DMW and the material was dried in an air oven at 50°C. The dried products were immersed in DMW to obtain small granules. They were converted to H-form by treating with 0.01 M HNO_3 for 24 h with occasional shaking intermittently replacing the supernatant liquid with fresh acid. The excess acid was removed after several washings with DMW and then dried at 50°C. Several particles size of materials were obtained by sieving and kept in desiccators [16].

Instrumentation

The applied adsorbents were thermally characterized by DTA-60 Shimadzu. and by FTIR analysis using BOMEM FTIR model MB 147, Canada. An elemental analyzer using atomic absorption spectrophotometer, AA6701F, Shimadzu. All samples and chemicals used in this work were weighted using an analytical balance of Bosch type having maximum sensitivity of 10^5 g and accuracy ±0.001/y. For the equilibrium experiments, a good mixing for the two phases was achieved using thermostatic shaker water bath of the type Julabo SW-20C, west Germany. All the pH values of different solutions were measured using an Orion digital pH meter research model 610A with microprocessor and have accuracy of ±0.02 units.

Elemental Composition

To determine the elemental composition of PATiW, the material was analyzed for Ti and W by X-ray fluorescence. Carbon, hydrogen and nitrogen contents of the material were determined by elemental analysis.

Chemical Solubility

The chemical solubility also plays an important role in the elucidation of properties of the ion-exchangers. Portions of 50 mg of composite and inorganic titanotungstate in the H^+ form were treated with 50 ml of varying concentration of acids, NaOH, and also in distilled water for 3 weeks with occasional shaking.

Ion-Exchange Capacity (IEC)

The IEC of polyaniline was determined by the repeated batch technique, by equilibrating 50 mg solid with 5 ml of 0.1 M cesium chloride solution on a shaker thermostat adjusted at 25°C ± 1°C and attain for equilibrium, then decontamination of solution was took place and saturation process was repeated until no further sorption. The solution was analyzed using atomic absorption in order to determine the amount of Cs^+ sorbet. The sorption (P) was calculated from the expressions:

$$P = \left(\frac{C_0 - C_t}{C_0} \right) \times 100$$

where C_0 is the initial concentration (mg/L) and C_t is the concentration at time t (mg/L) of metal ion in solution, V the volume (L) and m is the weight (g) of the adsorbent.

The sorption capacity was calculated using the following equation:

$$\text{capacity} = \frac{\%\text{uptake}}{100} \times C_0 \times V \times Z \quad \text{meq·g}^{-1}$$

where, C_0 is the initial concentration of solution, Z is the charge of adsorbed metal ion, V is the solution volume (ml), and m is the weight of the exchanger (g).

pH Titration

The Topp and Pepper method [17] was employed for pH titration studies of PATiW in solutions of alkali metal chlorides and their hydroxides. 200 mg was placed in a column that was fitted with glass wool at its bottom. A glass bottle containing 20 mL of 0.001 M HCl was placed below the column, and for determination of pH, a glass electrode was placed in the solution, then 75 mL of 0.1 M of NaOH was poured into the column. Titration was carried out, by passing the NaOH solution at a drop rate of about 1 mL/min. The pH of the solution was recorded until equilibrium was attained.

Distribution Studies

The distribution behavior of metal ions plays an important role in the determination of the material's selectivity. In certain practical applications, equilibrium is most conveniently expressed in terms of the distribution coefficients of the counter ions. The distribution coefficients (K_d) for different metal ions (Cs^+, Va^{5+}, Cr^{3+}, Zr^{4+}, As^{5+}, Mo^{6+}, Co^{2+}, Zn^{2+}, Cu^{2+} and Cd^{2+}) were determined by batch method as a function of pH. The K_d values of Cs^+ were determined by shaking 20 mg of each PATiW in H^+ form with 10 mL of HNO_3 solutions containing 10^{-4} mol·L^{-1} cesium ion solutions as chloride, where K_d values of other metal ions were determined by shaking 50 mg samples of PATiW in H^+ form with 5 mL of HNO_3 solutions containing 10^{-5} mol×L^{-1} metal ions as chloride in order to rise the sorption of this ions. After 24 h with intermittent shaking or continuous shaking for 6 h in a shaker at 25°C ± 1°C to attain equilibrium. The solutions were then filtered and metal ions were determined using AAS or ICP. The K_d values were calculated by the following equation;

$$K_d = \frac{(I-F)}{F} \frac{V}{m} \quad (\text{mL·g}^{-1})$$

where I is the initial amount of metal ion in the aqueous phase $(mg \times L^{-1})$, F is the final amount of metal ion in the aqueous phase $(mg \times L^{-1})$, V is the volume of the initial solution in ml and m is the dry mass of the ion exchanger in g.

Separation Factor

For the preferential uptake of the metal ion, the separation factor is determined in the separation of two metal ions. Separation factor α_B^A can be calculated as:

$$\alpha_B^A = \frac{K_d(A)}{K_d(B)}$$

where $K_d(A)$ and $K_d(B)$ are the distribution coefficient for the two competing species A and B in the ion-exchange system.

Sorption Isotherm

Batch adsorption studies of cesium ions was performed at different temperatures and neutral pH to obtain the equilibrium isotherms. A series of experiments were carried out by contacting a fixed amount of adsorbent 50 mg with 2.5 mL of Cs ion solution and have varying concentrations cover the range of 13 to 13290 $mg \times L^{-1}$ and agitated for a sufficiently time (~24 h) required to reach equilibrium. Then, adsorbent was decantation and the amount of metal ion retained in the adsorbent, q, was calculated using:

$$q = (c_0 - c_e) \frac{V}{m}$$

where C_0 and C_e are the initial and equilibrium concentration of Cs in aqueous solution.

Column Operation

A glass column of 1 cm diameter was used in this study. The column was packed of a fixed amount the composite and washed with distilled water and then all column studies were performed. The breakthrough curves (C/C_0 vs. volume) obtained for Cs^+ sorption onto PATiW at different bed depths 3.0 and 4.0 cm of 2.5 mL×min^{-1} flow rate and at 140 mg×L^{-1} of aqueous cesium solution. Also the breakthrough curves were carried out from acidic simulant (0.5 M HNO_3 + 0.1 M $NaNO_3$) and alkaline simulant (0.5 M NaOH + 0.1 M $NaNO_3$) solutions. The concentration of cesium in the feed solution was fixed at 13 mg×L^{-1} with 0.7 ml×min^{-1} flow rate and bed depth 1.0 cm.

The break-through percentage was calculated as

$$\text{Breakthrough} = \frac{C}{C_0}$$

where C and C_0 are the concentrations of cesium in the effluent and feed solution respectively.

The sorption capacity (q) of the PATiW was calculated by

$$q = \frac{V_{50} \times C_0}{W} \quad \text{mL·g}^{-1}$$

where V_{50} is the effluent volume corresponding to 50% breakthrough, C_0 is the concentration of cesium in the feed solution (mg×g^{-1}) and W is the mass of composite absorber (g).

Recover Cesium from Milk

Milk with 3%-fat solution was prepared with 10^{-2} M of cesium, this solution labeled with ^{134}Cs active. Kinetic studies are carried with added 5 ml of milk solution to 50 mg of PATiW and shaking in thermostated shaker water bath at 25°C, at interval time stopped shaker and withdrawn 1ml of milk solution for counting using a scintillation detector head (NaI) connected to scalar of the type SR-7 obtained by Nuclear Enterprises, USA and/or by Multichannel Analyzer Genie-200, spectroscopy system CANBERRA Industries INC (USA) allowed to equilibrate at room temperature. The Cs -ray activity in the tested milk was larger than background by at least 3 times.

RESULTS AND DISCUSSION

Titanium tungstate was found to be high stable, but have poor capacity for metal ions [18]. In an attempt to obtain materials with improved ion exchange properties and selective ion exchangers for the treatment of nuclear waste. So, PATiW has been synthesized by using this advanced class of inorganic ion-exchanger, which provided high selective andseparation for [134]Cs from metal ions. It was also noticed that the Cs^+ ion-exchange capacity of the composite material (1.82 meq×g^{-1}) [16].

Polyaniline can be easily synthesized chemically from acidic aqueous solutions. In this study, polyaniline gel was prepared by oxidative coupling using $K_2S_2O_8$ in an acidic aqueous medium at below 10°C as given below [19]:

$$4 \bigcirc - \overset{+}{NH_3} + 5S_2O_8^{2-}$$

Aniline $\downarrow H_2O, o-2\,C, 1\,h$

$$2\left[\bigcirc -NH-\bigcirc -NH-\right]^+ +12H^+ +10SO_4^{2-}$$

polyaniline

The effect of temperature on the reaction seems to be very pronounced. Aniline underwent oxidative coupling only below 10°C very effectively, leading to a good quantity of polyaniline with fairly good yield. The binding of polyaniline into the matrix of titanotungstate can be considered as:

TiW   Polyaniline

PATiW

The formation of inorganic precipitate TiW (X^-) was significantly affected by the pH of the mixture, and the most favorable pH of the mixture was 6.5. The preparation of the inorganic precipitate at pH lower or higher than 6.5 lead to decrease in yield and in ion-exchange capacity of the material.

The weight percent composition of the material was found to be Ti, 15.15%; W, 44.6%; C, 19.1%; H, 2.3%; N, 3.7%; and O, 15.0%. The corresponding molar ratio of Ti, W, C, H, N, and O in the material was estimated as 1.34:1.0:6.19:9.1:1.015:3.6, which can suggest the following formula of the material:

$$[(TiO_2)(WO_3)(-C_6H_4NH-)] \cdot nH_2O.$$

Assuming that only the external water molecules (n) are lost at 130°C, the 10% weight loss of mass represented by TGA curve must be due to the loss of nH_2O from the above structure. The value of (n) can be calculated using Alberti's equation:

$$18n = X(M + 18n)/100$$

where X is the percent weight loss (10%) of the exchanger by heating up to 130°C and (M + 18n) is the molecular weight of the material. The calculations indicated that ~2.46 per molecule of the cation-exchanger. Based on the literature data [19,20] the structural water molecules may play an important rule as exchange sites.

The solubility experiments showed that the composite and inorganic ion exchangers have good chemical stability (Table 1). As the results indicated that the materials were resistant to 6 M HNO_3 and 6 M HCl. The solubility in the HNO_3 is higher than in the HCl and very feeble dissolution was observed in the alkaline medium. There is no chemical dissolution in DMW. It was observed that the chemical solubility of composite is slightly increased than the inorganic material, due to the presence of polyaniline which can dissolute into the solution [16]. Despite this increased in chemical solubility of PATiW than TiW but the mechanical and granular properties of PATiW are higher than TiW [19].

Table 1: Chemical solubilty of TiW and PATiW in various solvent

Concentrations of Solvent used (50 all)	Amount dissolved (mg/L)			
	HCl		HNO_3	
	TiW	PATiW	TiW	PATiW
0.1 M HCl	4.0	6.0	12.0	20.0
0.5 M HCl	10.0	16.0	18.0	24.0
1,M HCl	24.0	34.0	30.0	38.0
2,M HCl	30.0	38.0	40.0	46.0
3M HCl	40.0	44.0	62.0	96.0
4,M HCl	56.0	82.0	86.0	114.0
O.M HCl	72.0	94.0	102.0	124.0
k.M HC1	80.0	98.0	118.0	140
water	0.00	0.00		
0.1114 NaOH	0.2	0.4		
1,M NaOH	6.0	10.0		

The pH-titration curve of the PATiW shows only one inflexion point indicating that the PATiW behaves as monofunctional. The pH-titration curve (Figure 1) showed slow increase when NaOH was added 0.30 - 1.00 mmol and rabid increase when NaOH added 1.00 - 1.95 mmol. This composite exchanger may be a strong acid cationexchanger because the pH-titration curve usually showed a step edge at 1.95 $mmol \cdot g^{-1}$. This means that the H^+ ions on the hybrid cation-exchanger were depleted and replaced with Na^+ ions at that point and the number of H^+ sites were equivalent to the same amount of NaOH, i.e. the strong acidic groups (H^+) of the composite cationexchanger are completely converted to the Na^+ form [16]. Thus, theoretical ion-exchange capacity of this hybrid cation-exchanger may be considered as 1.95 mmol·g^{-1}. After that point, in the region when more NaOH added, the equilibrium pH further increases but more slowly. This slow increase of pH-titration curve after 1.95 mmol·g^{-1} implies due to surface precipitation other than conventional ion exchange or surface adsorption.

The ability of PATiW composite for exchange of cesium is significantly affected by the composition ratio of TiW. The preparation of PATiW with ratio (1:1:1) of poly aniline:titanium:tungstate give the percent absorption (99%, 2.6%, 2.6%) for Cs^+, Co^{2+} and Eu^{3+} respectively at 10^{-4} M where, The preparation of PATiW with ratio (2:1:1) of polyaniline:titanate:tungstate give the percent sorption (78%, 7%, 17%) for Cs^+, Co^{2+} and Eu^{3+} respectively this means that the selectivity is decreased than the first ratio, this characterization is showing as in Table 2.

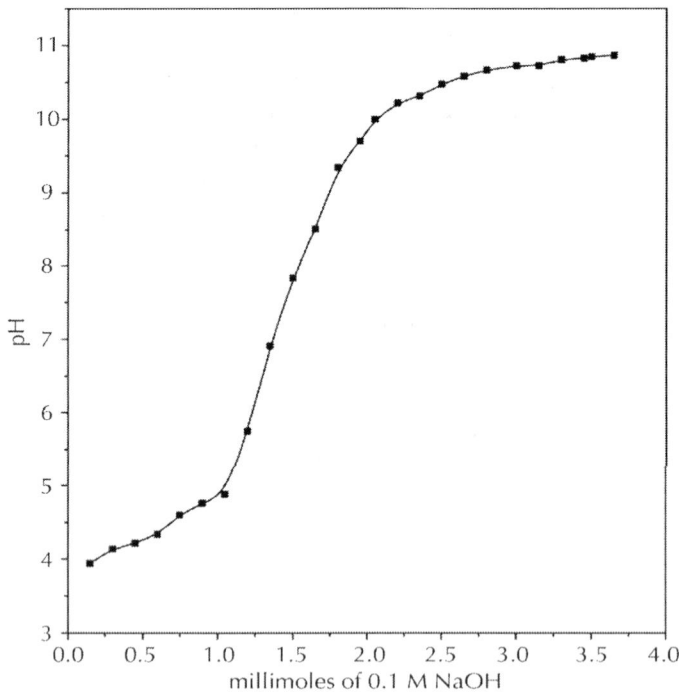

Figure 1: The pH-titration curve of PATiW with 0.1 M NaOH.

Also the characterization and ability of PATiW for sorption of cesium ions is significantly affected by the concentration of the composition of TiW where by preparing PATiW by adding polyaniline to 0.1 M for each of titanate and tungastate, the results of the product is not selectivity for cesium and soluble in dilute acids.

However, TiW exhibited high granulometric and mechanical properties, showing a good reproducible behavior as is evident from the fact that these materials obtained from various batches did not show any appreciable deviation in their ion-exchange capacities.

Figure 2 shows the ion exchange capacity of PATiW for Cs^+ as a function of pH. It was found that the capacity increase by increasing the pH value. This is may be attributed to, with increasing the pH of the solution the $[H]^+$ in solution is decrease which facilitate the released of H^+ from the exchanger to solution. So the % uptake values were increased and thus the capacity was increased [16].

Titanium tungstate was found to be high stable, but have poor capacity for metal ions [21]. In this work, an attempt to obtain materials with improved ion exchange properties and have high efficiency for the treatment of nuclear waste. So, PATiW has been synthesized to represent this advanced class of inorganic ion-exchanger, which provided high selective and separation performance for cesium ions of nuclear waste. It was also noticed that the Cs^+ ion-exchange capacity of the composite material PATiW (1.82 meq$\times$g^{-1}) is higher as compared to inorganic ion-exchanger TiW (0.6 meq$\times$g^{-1}) [22], magneso-silicate (0.57 meq$\times$g^{-1}) and magnesium aluminosilicate (0.77 meq$\times$g^{-1}) [23]. Also it higher than the saturation capacity (meq$\times$g^{-1}) which decrease in the order; Gd^{3+} (1.63) > Eu^{3+} (1.41) > Ce^{3+} (1.34) meq$\times$g^{-1} on titanium (IV) antimonite [24]. The obtained ion-exchange capacity of Cs^+ on PATiW is higher than the ion-exchange capacity of Cs^+ on lithium zirconium silicate [25]. The obtained ion-exchange capacity of Cs^+ on PATiW is higher than the ion-exchange capacity of Cs^+ on lithium zirconium silicate [24]. The ion exchange capacity on PATiW is higher than that for Cs^+ on magnesium and cerium titano-antimonates in aqueous (0.7 meq$\times$g^{-1}), in 25% methnol (0.74 meq$\times$g^{-1}) and in 25% ethanol (0.8 meq$\times$g^{-1}) [25].

Table 2: Factors affecting on preparation of PATiW

Sample	Mixing volume ratio(v/v)		Mixing volume ratio(v/v) of aniline	%Aniline	Appearance after drying	Percent Sorpotion		
	$TiCl_4$	Na_2Wo_2				Cs^+	Co^{2+}	Eu^{3+}
PATiW-1	1(1 M)	1(1 M)	1	10	Black shiny granules	99	2.6	2.6
PATiW-2	1(0.1 M)	1(0.1 M)	2	10	Black shiny granules	78	7	17

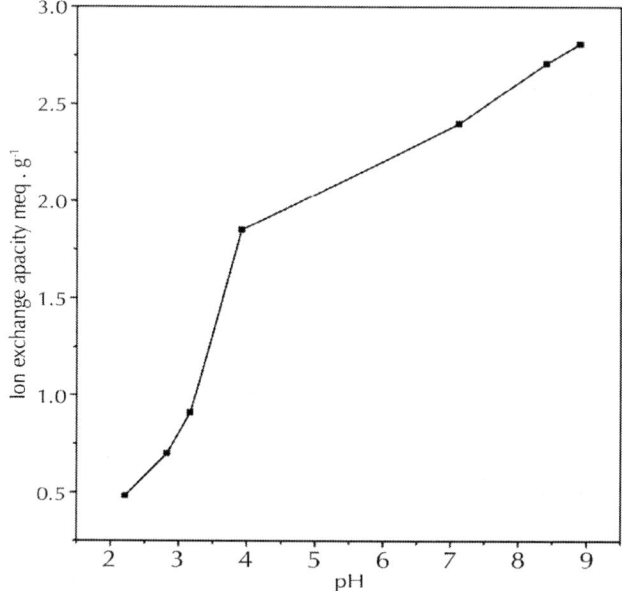

Figure 2: Plots of capacity against pH for exchange of Cs⁺ on PATiW at 0.1 M and 25°C ± 1°C.

The mechanism of the exchange of Cs⁺ with PATiW can be considered as the following scheme:

$$R-H^+ + Cs^+ \rightleftharpoons R-Cs + H^+$$

Exchanger phase   Solution phase   Exchanger phase   Solution phase

$$Cs^+ \left[\langle\bigcirc\rangle\text{-NH-}\langle\bigcirc\rangle\text{-NH-} \right]^{+} \left[(TiO_2)(WO_3) \right]^{-} 2.46H_2O$$

$$\updownarrow$$

$$\left[\langle\bigcirc\rangle\text{-NH-}\langle\bigcirc\rangle\text{-}\overset{H}{N}\text{-} \right]^{+} \left[(TiO_2)(WO_3) \right]^{-} 2.46H_2O \; Cs^+$$

where the negative charge of groups is compensated by Cs^+. Possible associations between Cs^+ and amine groups are indicated by dash-lines ovals.

Characterization of PATiWare described in details using IR, XRD spectrum and TGA-DTA analysis [26].

The distribution coefficient is often a proper quantity to express the distribution of an ion between the exchanger and the solution phase. This is especially true when the exchanging ion is present in the trace concentration, since the ionic composition does not practically change at macro levels in trace ion exchange. So the K_d values of Cs^+ were determined by shaking 40 mg of PATiW sample with 20 mL of HNO_3 solutions containing 10^{-4} mol·L^{-1} cesium ion to give different pH at different reaction temperatures (25°C, 45°C and 60°C ± 1°C).

The log K_d for Cs^+ on PATiW were determined at 25°C, 45°C and 60°C ± 1°C as afunction of pH using different concentrations of HNO_3 as shown in Figure 3. The preliminary studies indicated that, the time of equilibrium for the exchange of Cs^+ with H^+ form PATiW was attained within 24 h (sufficient to attain the equilibrium). From the results shown in Figure 3, it can be found that the distribution coefficients for Cs^+ on PATiW was increased with increasing the pH of solutions, this trend is an obvious phenomenon and was observed [24,25]. A linear relationship with a slope smaller than the valences of the Cs^+ was obtained. Analysis the data shown in Figure 3 indicated that the ion exchange reaction deviated from the ideal process. In addition, it was found that the K_d values increased with increasing the temperature. The non-ideality may due to a different mechanism such as physical adsorption, chemical reaction or other effects, which takes place besides the ion exchange process [16, 27].

The work was directed towards the PATiW as a more efficient exchange material for Cs^+ separation. In order to find out the potentiality of the composite cation exchanger in the separation of metal ions, distribution studies for 10 metal ions were performed at different pH and the values obtained for distribution coefficients are given in **Table 3**. The distribution coefficients (K_d) for different metal ions (Cs^+, Va^{5+}, Cr^{3+}, Zr^{4+}, As^{5+}, Mo^{6+}, Co^{2+}, Zn^{2+}, Cu^{2+} and Cd^{2+}) on PATiW were determined by batch method as a function of pH. In this metal ions the K_d values were determined by shaking 50 mg of PATiW with 5 mL of HNO_3 solutions containing 10^{-5} mol×L^{-1} metal ions.

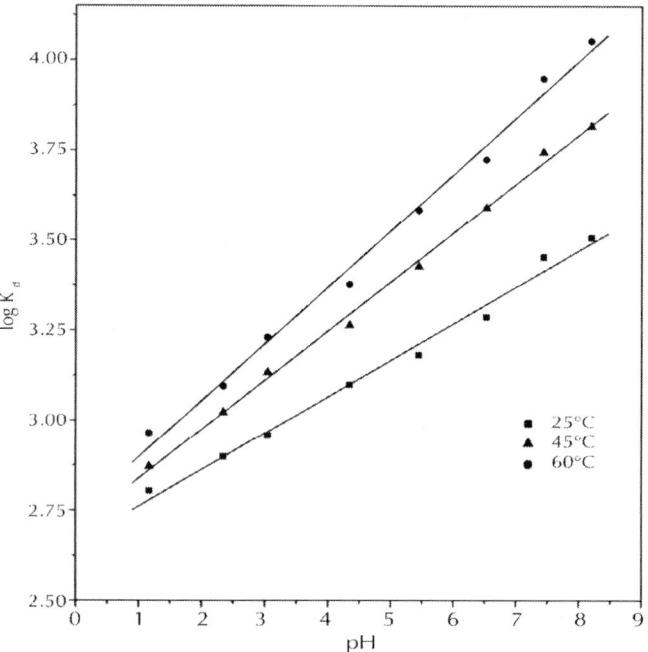

Figure 3: Plots of log K_d against pH for exchange of Cs^+ on PATiW at different reaction temperatures.

The distribution studies showed that the PATiW was found to be the highly selective for cesium while the other metal ions (Va^{5+}, Cr^{3+}, Zr^{4+}, As^{5+}, Mo^{6+}, Co^{2+}, Zn^{2+}, Cu^{2+} and Cd^{2+}) were poorly sorbed on

PATiW as in Figure 4. The high uptake of Cs$^+$ demonstrates not only the ion-exchange properties but also the adsorption and ionsieve characteristics of the cation-exchanger.

The effect of the size and charge of the exchanging ions on K_d values was also observed on PATiW. The K_d values show a decreasing trend in this order Cs$^+$ >>> Zr^{4+} > Mo^{6+} > Va^{5+} > As^{3+} > Cr^{3+} > Co^{2+} > Cu^{2+} > Zn^{2+} > Cd^{2+} as in Figure 4.

This sequence is in accordance with the unhydrated radii of the exchanging ions. Ions with the smaller unhydrated radii easily enter the pores of the exchanger, which results in higher adsorption [16]. According to [28] the attraction between cations and anions in ionic crystals obey Coulomb's law on the demands for ions of equal charge, a small ion will be attracted either to a greater force or held more tightly than a larger ion as in **Table 3**.

Therefore, the K_d value should increase with decreasing unhydrated radii and increase with electricpotential. On the basis of distribution studies, the most promising property of the material was found to be the high selectivity towards Cs$^+$, which is a major polluting metal in the fission products in radioactive wastes and thus is due to the spacing of the lattice is believed to correspond very closely to the ionic radius of unhydrated Cs$^+$ [28] and, thus, is responsible for its selectivity. The selectivity order; Cs$^+$ > Eu^{3+} > Co^{2+} onto phosphoric acid activated silico-antimonate, the unhydrated ionic radius was regarded to play a major role affecting on many of these selectivity behavior [29]. The same behavior sequence is similar on resorcinol-formaldehyde (R-F) and zirconylmolybdopyrophosphate (ZMPP) [30], where the selectivity order on R-F and ZMPP samples is Cs$^+$ > Co^{2+} > Eu^{3+} > Zn^{2+}.

The distribution coefficient for Cs$^+$ is 14,000 at pH 8.5 on PATiW is greater than the values of K_d for Cs$^+$ on stannic molybdophosphate ion exchanger at the same pH [31] and on zirconium vanadate at the same pH [32]. The K_d values for Co^{2+} and Zn^{2+} on PATiW are lower than the K_d values for Co^{2+} and Zn^{2+} on Polyaniline Sn(IV) arsenophosphate and Nylon-6,6, Zr(IV) phosphate [33]. On the reverse of that, the K_d values for Cd^{2+} on PATiW are lower than the K_d values for Cd^{2+} on polyaniline Ce(IV) molybdate [34]. The K_d values for Cu^{2+} on PATiW are lower than the K_d values for Cu^{2+} on acrylamide stannic silicomolybdate [35].

Return to the previous results, regarding that, the studied systems offer very good selectivity for cesium ion. The separation capability of the material has been demonstrated by achieving some important binary separations such as Cs-Co, Cs-Zn, Cs-Cd, Cs-Cu, Cs-Cr, Cs-AS, Cs-Zr, Cs-V, and Cs-Mo as in Table 3.

In order to evaluate the maximum metal sorption capacity of PATiW, the sorbent was contacted with varying concentrations of Cs^+ (13 - 13,290 mg×g^{-1}) until equilibrium was reached. Cs^+removing on PATiW was increase with increasing ion concentration in solution until it reached the maximum capacity of PATiW at different reaction temperatures. Equations often used to describe the experimental isotherm data are those developed by Freundlich [36] and by Langmuir [37].

The empirical model of Freundlich equation can be applied to non-ideal sorption on heterogeneous surfaces as well as multilayer sorption and is expresses by the following equation:

$$q_e = K_f C_e^{1/n}$$

where q_e is the amount of metal ions sorbed per unit weight of PATiWin equilibrium (mg/g), C_e is the equilibrium liquid phase concentration (mg/L), $K_f$the Freundlich constant indicative of the relative sorption capacity (mg/L) and 1/n is the heterogeneity factor indicative of the intensity of the sorption process. A linear form of the Freundlich expression can be obtained by:

$$\ln q_e = \ln K_f + \frac{1}{n} \ln C_e$$

The Freundlich constants are empirical constants which depend on several environmental factors. The magnitude of the exponent 1/n gives an indication of the adequacy and capacity of the adsorbent/adsorbate system [38]. In most cases, an exponent between 1 and 10 shows beneficial adsorption. The value of n ranges between 0 and 1, and indicates the degree of non-linearity between solution concentration and adsorption as follows [39]: if the value of n is equal to unity, the adsorption is linear; if the value is below unity, this implies that the adsorption process is chemical; if the value is above unity, adsorption is a favorable physical process; the more heterogeneous the surface, the closer n value is to 0 [40].

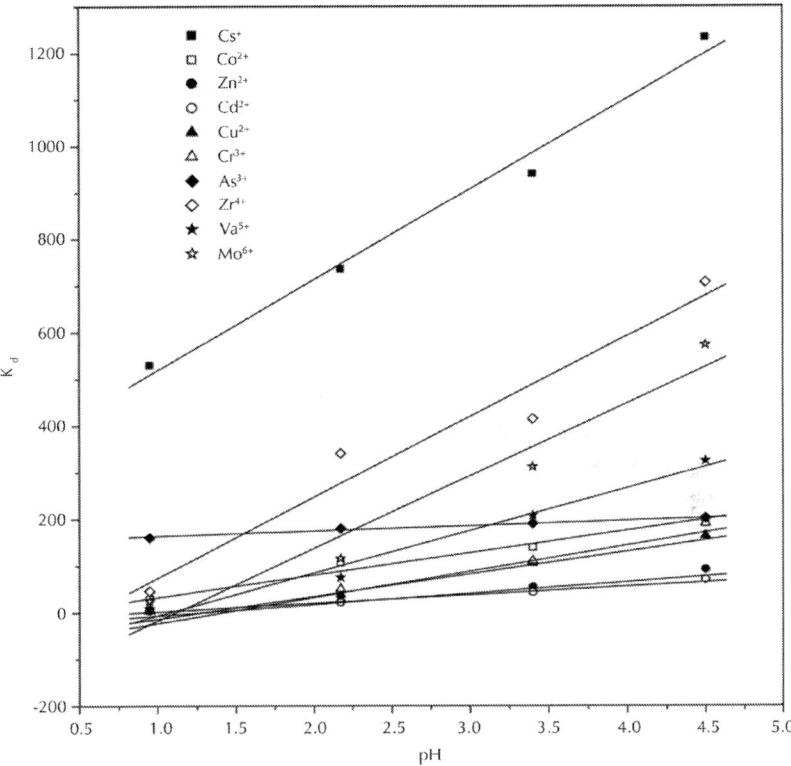

Figure 4: Plots of K_d against pH for exchange of Cs^+, Co^{2+}, Zn^{2+}, Cd^{2+}, Cu^{2+}, Cr^{3+}, As^{5+}, Zr^{4+}, Va^{5+} and Mo^{6+} at 25°C ± 1°C on PATiW at 10^{-4}M of Cs^+ and 10^{-5} M for other ions.

Table 3: K_d values of Cs^+, Co^{2+}, Zn^{2+}, Cd^{2+}, Cu^{2+}, Cr^{3+}, As^{5+}, Zr^{4+}, Va^{5+}, Mo^{6+} and separationfactors ($\alpha_B^{Cs^+}$) of Cs^+ from other metal ions at different concentrations of HNO_3 on PATiW 25°C ± 1°C

Mental ions	Cs^+	Co^{2+}	Zn^+	Cd^{2+}	Cu^{2+}	Cr^{2+}	A^{5-}	Zr^{4+}	V^{5+}	Mo^{6+}
HNO_3, M										
Water	12562	199.4	91.9	69.9	163.5	199.4	201.2	733	324.1	572
		6.3	13.6	17.97	7.68	63	6.2	1.7	3.87	2.2
10^{-3}M	907.65	119.6	35.7	42.3	1053	91,	190.5	415	207	312
		7.5	25.4	21.45	8.6	9.9	4.76	2.18	438	2.9
		139.8	10.5	15.9	11.5	7.1	184.9	340.5	74.8	115.5
10^{-2}M	750.31	5.4	71.4	47.18	65.24	105.7	4.05	2.2	10.0	6.5
		11.11	3.9	7.6	3.6	4.2	157.8	46.4	7.6	19.9
10^{-1}M	661.17	59.51	169.5	86.99	183.7	157.4	4.2	14.24	86.9	33.22

The fit of data to Freundlich isotherm indicates the heterogeneity of the sorbent surface. The linear plot of $\ln q_e$ versus $\ln C_e$ (Figure 5) for polyaniline titanotungstste shows that the adsorption obeys to the Freundlich model.

Figure 5 for PATiW show the sorption of Cs^+ obey Freundlich isotherm over the entire range of sorption concentration studied. Similar results is found for sorption of Cs^+ with Freundlich isotherm onto polyacylamide cerium titanate [41].

The numerical values of the constants n and K_f are computed from the slope and the intercepts, by means of a linear least square fitting method, and also given in Table 4. It can be seen from these data that the Freundlich intensity constants (n) are greater than unity for PATiW. This has physicochemical significance with reference to the qualitative characteristics of the isotherms, as well as to the interactions between metal ions and both adsorbents.

In our case, n > 1 for Cs^+, the PATiW shows an increase tendency for sorption with increasing solid phase concentration. This should be attributed to the fact that with progressive surface coverage of adsorbent, the attractive forces between the metal ions such as Vander Waals forces, increases more rapidly than the repulsiveforces, exemplified by short-range electronic or long range Coulombic dipole repulsion, and consequently, the metal ions manifest a stronger tendency to bind to the adsorbent site [42].

Langmuir sorption isotherm model described the monolayer coverage of the sorption surfaces and assumes that sorption occurs on a structurally homogeneous adsorbent and all the sorption sites are energetically identical.

The Langmuir model is probably the best known and most widely applied sorption isotherm. It may be represented as follows:

$$\frac{C_e}{q_e} = \frac{1}{q_o b} + \frac{C_e}{q_o}$$

where q_e is the amount of cesium ions sorbed per unit weight of PATiW (mg/g), C_e the equilibrium concentration of the cesium ions in the equilibrium solution (mg/L), q_o the maximum adsorption capacity corresponding to complete monolayer coverage on the surface (mg/g), and b is the Langmuir constant (L/mg) related to the ($b \propto e^{-\Delta G/RT}$) free energy of adsorption.

The linear plot of (C_e/q_e) versus C_e give straight lines for Cs^+ sorbed onto both adsorbent, as presented in Figure 6, confirming that this expression is indeed a reasonable representation of chemisorptions isotherm.

The numerical value of constants q_o and b evaluated form the slopes and intercepts of each plot are given in Table 4. The value of saturation capacity q_o corresponds to the monolayer coverage and defines the total capacity of the adsorbent for a specific metal ion. As it can be seen from Table 4, the monolayer sorption capacity (q_o) values of composite towards Cs^+ at 25 45 and 60 ± 1 are 227.79, 292.39 and 332.23 respectively.

The Langmuir constants q_o and b for Cs^+ sorbed onto both adsorbent, increased with temperature showing that the sorption capacity and intensity of sorption are enhanced at higher temperatures.

This increase in sorption capacity with temperature suggested that the active surface available

for sorption has increased with temperature and that adsorption capacity (q_o) increases with the increase in temperature, this indicating that the process was endothermic in nature.

It was observed that the equilibrium adsorption data indicating the favorable Langmuir's sorption isotherms of Cs^+ onto adsorbent.

Conformation of the experimental data into Langmuir isotherm model indicates the homogeneous nature of PATiW as surface, i.e. each Cs^+/ PATiW adsorption has equal adsorption activation energy and demonstrates the formation of mono layer coverage of Cs^+ on the outer surface of PATiW.

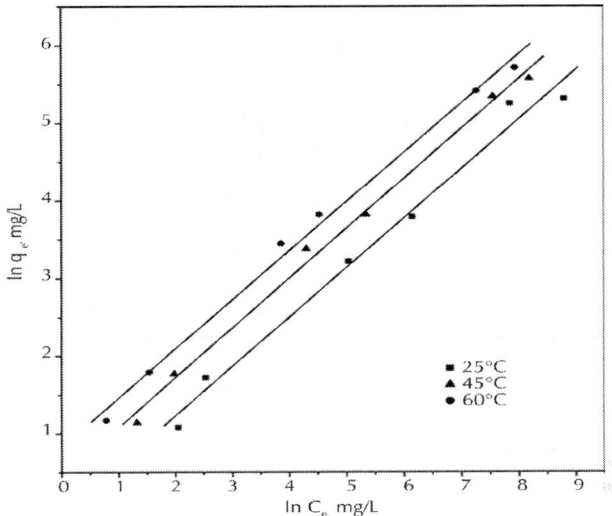

Figure 5: Freundlich isotherm plots for sorption of Cs+ onto PATiW at diffrent reaction temperatures.

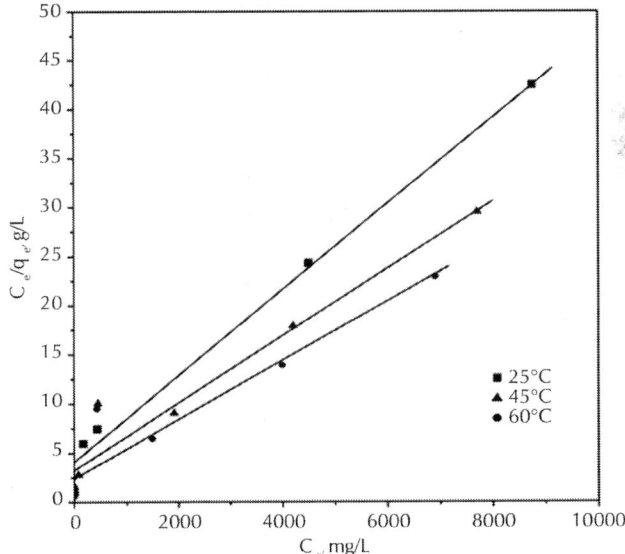

Figure 6: Langmiur isotherm plots for sorption of Cs+ onto PATiW at diffrent reaction temperatures.

Table 4: Freundlich and Langmuir isotherm parameters for the sorption of Cs^+ onto PATiW

Adsorbent	Temperature	Freundlish model parameters				Langmiur model parameter		
		n	k_f(mg/g)	R^2	Q^o(mg/g)	B(L/mg/g) $\times 10^{-3}$	R^2	R_L
PATiW	25°C	1.56	0.96	0.993	227.79	1.07	0.992	0.067
	45°C	1.57	1.55	0.997	292.39	1.09	0.968	0.069
	60°C	1.58	2.29	0.997	332.23	1.26	0.953	0.058

One of the essential characteristics of the Langmuir model could be expressed by dimensionless constant called equilibrium parameters R_L [43]:

$$R_L = \frac{1}{1+bC_0}$$

where C_0 is the highest initial concentration of adsorbate (mg/L) and b (L/mg) is Langmuir constant. The parameter R_L indicates the nature of shape of the isotherm accordingly: $R_L > 1$ unfavorable adsorption; $0 < R_L < 1$ favorable adsorption; $R_L = 0$ irreversible adsorption; $R_L = 1$ linear adsorption All the R_L values (Table 4) were found to be less than 1 and greater than 0 indicating the favorable sorption isotherms of adsorbent Cs^+ onto PATiW and the used these adsorbents are optimum for removal of Cs^+ from waste solutions. The values of R_L were determined at different temperatures 25°C, 45°C and 60°C ± 1°C over the broad concentration range and the results are shown in Figure 7 for PATiW. All the R_L values were found to be less than one and greater than zero indicating the favourable adsorption of Cs^+ onto PATiW.

The detailed analysis of the R^2 values showed that the Freundlich model fit the adsorption data better than the Langmuir model at different temperatures. Freundlich sorption isotherm does not predict any saturation of the solid surface thus envisages infinite surface coverage mathematically [44]. Which indicates that Cs^+ sorbed on PATiW as a monolayer deposition of adsorbate on localized sites followed by a multilayer sorption with interaction between sorbed molecules that having heterogeneous energy distribution, accompanied by interaction be tween the adsorbed molecules. The same result was found for physically sorbed of Cs^+ on magneso-silicate and magnesium aluminosilicate [45]. The theoretical capacity (q_o) of Cs^+ on PATiW were calculated as 227.8 and 181.8 $mg \times g^{-1}$, respectively, against 205.9 and 156.15 $mg \times g^{-1}$ found experimentally indicating that this sorption is not completely match with Langmiure isotherm.

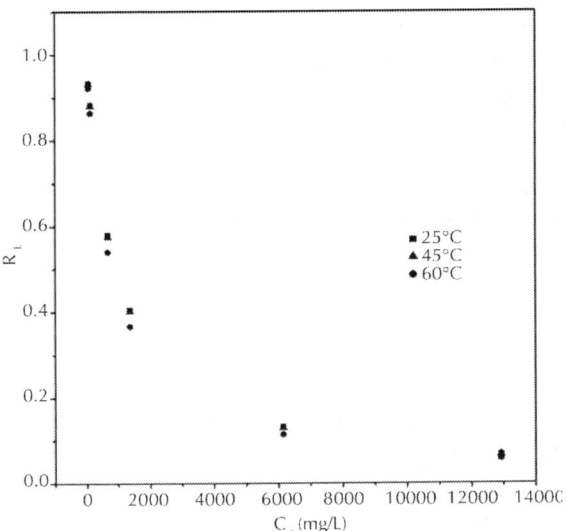

Figure 7: Plots of separation factor, R_L, against initial concentration, Co, for sorption of Cs^+ onto PATiW at different reaction temperatures.

Breakthrough curves of PATiW for the conditions stated previously are shown in Figures 8 and 9. Batch experimental data are often difficult to apply directly to the fixed bed sorption column because isotherms are unable to give accurate data for scale up since a flow in the column is not at equilibrium. Fixed bed column sorption experiments were carried out to study the sorption dynamics. The fixed bed column operation allows more efficient utilization of the sorption capacity than the batch process. The shape of the breakthrough curve and the time for the breakthrough appearance are the predominant factors for determining the operation and the dynamic response of the sorption column [16]. The general position of the breakthrough curve along the volume/time axis depends on the capacity of the column with respect to bed height, the feed concentration and flow rate [46].

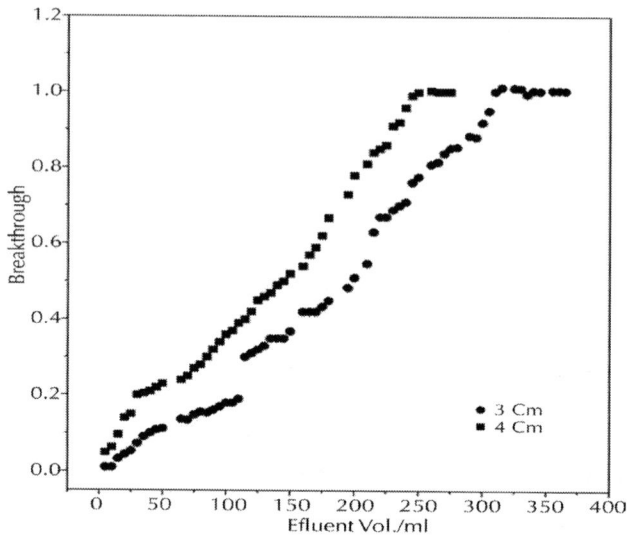

Figure 8: Performance of PATiW column of cesium seperation from neutral solutions at different bed depth, 140 mg×L^{-1} and flow rate 2.5 ml×min^{-1}.

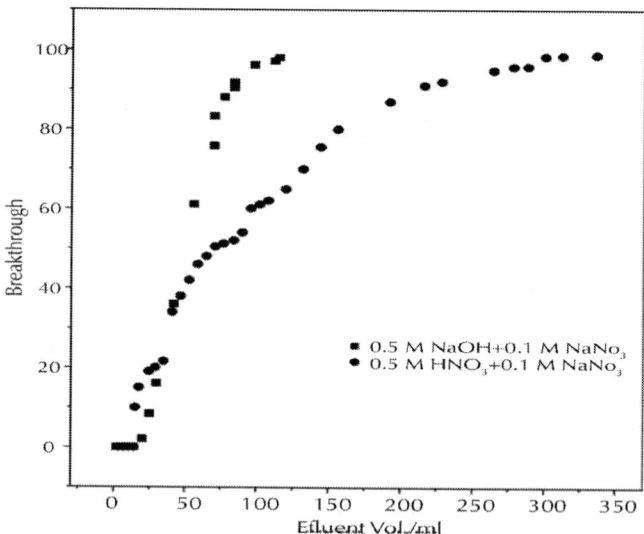

Figure 9: Performance of PATiW column of cesium separation from alkaline and acidic simulant solutions at bed depth 1 cm, 13 mg×L^{-1} and flow rate 0.7 ml×min^{-1}.

The breakthrough curves (C/C_0 vs volume) obtained for Cs^+ sorption onto PATiW at different bed depths (3.0, and 4.0 cm) for a constant linear flow rate of 2.5 mL/min and at 140 mg/L of neutral aqueous cesium concentration are shown in Figure 8.

It can be observed a similar behavior in each curve and a tendency to follow an S shape which is characteristic of an ideal sorption. The results indicate that the volume of breakthrough varies with bed depth. The bed capacity and the percent removal (column performance) for Cs^+ increased with increasing bed height, as more binding sites were available for sorption. The increase in the ion sorption with bed depth was due to the increase in the sorbent doses in larger beds, which provided greater sorption sites for Cs^+. It was found that the breakthrough capacities for Cs^+ onto PATiW at different bed depths 3.0 and 4.0 cm are 5.1 and 6.5 mg×g^{-1} respectively. Which there are more than the breakthrough capacity 0.67 mg×g^{-1} for Cs^+ onto phosphoric acid activated silicoantimonate [47].

The rate determining step can be inferred from a stopflow test, in which the flow is halted and restarted during column loading. The behavior of C/C_0 after the column is restarted provides information about the mass transfer mechanism. If the exchange rate is controlled by diffusion in the particle phase, diffusion of Cs^+ within the particles continues even after flow is stopped. Highly concentrated cesium on the outer layers of the particles will diffuse toward particle centers, there by leveling the concentration gradient in the particle and reducing the C/C_0 on the surface. The result is a decrease in C/C_0 when the column is restarted. As the run continues, the concentration gradients in the particles are reestablished and the breakthrough curve will slowly approach the shape it would have had without interruption [48].

In each of the PATiW column test where flow was interrupted (Figure 8) there was a significant decrease in C/C_0 when the operation was restarted and it took approximately 50 - 150 bed volumes for the curve shape to be re-established. This phenomenon is indicative of a particle diffusion controlled system. This stop flow test is also analogous to the batch interruption test reported by [49] in which the sorbent particles are removed from the solution for a brief period of time and then re-immersed [16]. The interruption gives time for concentration gradients in the solid phase to level out. Then, when the particles are re-immersed, the exchange rate is temporarily faster. This can be seen as a momentary increase in the fractional attainment of equilibrium in the time following re-immersion.

The breakthrough curves for the separation of cesium from acid solution (0.5 M HNO$_3$ + 0.1 M NaNO$_3$) and from alkaline simulant solution (0.5 M NaOH + 0.1 M NaNO$_3$) using PATiW columns at bed depth 1 cm, flow rate of 0.7 ml×min^{-1} and 13 mg×g^{-1} of cesium chloride were represented in Figure 9. In this figure the breakthrough of cesium begins earlier with respect acid stimulant solution (at 22 ml) than with respect to alkaline solution begin very early (at 10 ml) and the sorption capacity was found to be 2.32 and 0.66 mg×g^{-1} for acid and alkaline columns, respectively. This means that PATiW can be applied to remove radiocesium from acidic solutions where most of the inorganic ion exchangers [50,51] such as lithium titanate, tin silicate and titanium-ferrocyanides, exhibit very low ion exchange efficiency in the high acidic media. Where for alkaline solution the breakthrough begins very early with capacity very small.

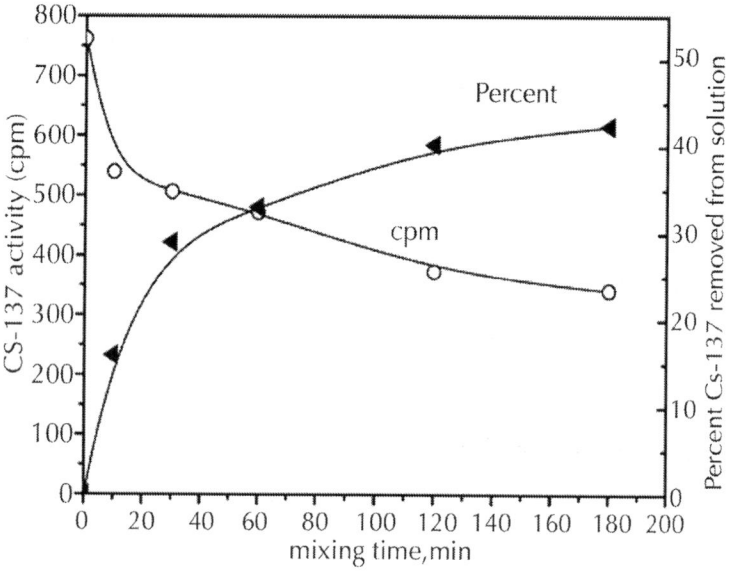

Figure 10: Partitioning of cesium-134 as a function of mixing time.

This system offer very good selectivity for cesium sorption, since if the break-through was studied where the feed solution is a simulated active waste solution containing different elements of various valence states such as Cs^+, Va^{5+}, Cr^{3+}, Zr^{4+}, As^{5+}, Mo^{6+}, Co^{2+}, Zn^{2+}, Cu^{2+} and Cd^{2+}. So all these elements could be separated in the first few milliliters of the effluent solution while cesium is retained completely on the column beds.

The milk cesium-134 activity was measured over a 3-h period to determine the equilibrium time. Fat separation was not observed except for long mixing time (>3 h) [52]. The uptake of cesium-134 and the decaying of cesium activity are shown as a function of mixing time in Figure 10. The reaction half-life was 30 min, and by 60 min the reaction was 80% complete. The reaction reached at equilibrium at 2 h [16].

CONCLUSIONS

A cesium selective composite cation exchanger PATiWhaving good ion-exchange capacity (1.8) have been prepared successfully. Cesium ion sorption was fit best to the Freundlich isotherm model. The breakthrough capacity was different with the bed depths and the solution that used. The PATiWcan be used to removal cesium ion from aqueous, high acid solutions and milk.

REFERENCES

1.   IAEA, "Remediation of Areas Contaminated by Past Activities and Accidents, Safety Requirements," Safety Standards Series No. WSR-3, STI/PUB/1176, IAEA, Vienna, 2003.

2.   P. J. Coughtrey and M. C. Thorne, "Radionuclide Distribution and Transport in Terrestrial and Aquatic Ecosystems," A. A. Balkema, Rotterdam, 1983.

3.   A. Clearfield, "Inorganic Ion Exchange Materials," CRC Press, Boca Raton, 1982.

4.   A. A. Khan and M. M. Alam, "New and Novel OrganicInorganic Type Crystalline 'Polypyrrolel/Polyantimonic Acid' Composite System: Preparation, Characterization and Analytical Applications as a Cation-Exchange Material and Hg(II) Ion-Selective Membrane Electrode," Analytica Chimica Acta, Vol. 504, 2004, pp. 253-264.doi:10.1016/j.aca.2003.10.054

5.   A. A. Khan, "Inamuddin, Preparation, Physico-Chemical Characterization, Analytical Applications and Electrical Conductivity Measurement Studies of an 'Organic-InorGanic' Composite Cation-Exchanger: Polyaniline Sn(IV) Phosphate," Reactive & Functional Polymers, Vol. 66, 2006, pp. 1649-1663. doi:10.1016/j.reactfunctpolym.2006.06.007

6.   A. A. Khan and M. M. Alam, "Determination and Separation of Pb^{2+} from Aqueous Solutions Using a Fibrous Type Organic-Inorganic Hybrid Cation-Exchange Material: Polypyrrole Thorium(IV) Phosphate," Reactive & Functional Polymers, Vol. 63, No. 2, 2005, pp. 119-133. doi:10.1016/j.reactfunctpolym.2005.02.001

7.   A. A. Khan and M. M. Alam, "Synthesis, Characterization and Analytical Applications of a New and Novel 'Organic-Inorganic' Composite Material as a Cation Exchanger and Cd(II) Ion-Selective Membrane Electrode: Polyaniline Sn(IV) Tungstoarsenate," Reactive & Functional Polymers, Vol. 55, No. 3, 2003, pp. 277-290. doi:10.1016/S1381-5148(03)00018-X

8.   R. Niwas, A. A. Khan and K. G. Varshney, "Synthesis and Ion Exchange Behaviour of Polyaniline Sn(IV) Arsenophosphate: A Polymeric Inorganic Ion Exchanger," Colloids and Surfaces A, Vol. 150, 1999, pp. 7-14. doi:10.1016/S0927-7757(98)00843-7

9.   K. G. Varshney, N. Tayal and U. Gupta, "Acrylonitrile Based Cerium (IV) Phosphate as a New Mercury Selective Fibrous Ion-Exchanger: Synthesis, Characterization and Analytical Applications," Colloids and Surfaces A, Vol. 145, No. 2-3, 1998, pp. 71-81.doi:10.1016/S0927-7757(98)00657-8

10.   G. Alberti, M. Casciola, C. Dionigi and R. Vivani, Proceedings of International Conference on Ion-Exchange, ICIE'95, Takamtsu, 1995.

11.   A. Mardan, R. Ajaz, A. Mehmood and S. M. Raza, "Preparation of Silica Potassium Cobalt Hexacyanoferrate Composite Ion Exchanger and Its Uptake Behavior for Cesium," Separation and Purification Technology, Vol. 16, No. 2, 1999, pp. 147-158. doi:10.1016/S1383-5866(98)00121-X

12.   Z. Alam and S. A. Nabi, "Synthesis and Characterization of a Thermally Stable Strongly Acidic Cd(II) Ion Selective Composite Cation-Exchanger: Polyaniline Ce(IV) Molybdate," Desalination, Vol. 250, No. 2, 010, pp. 515-522.

13. Y. A. Ismail, "Synthesis and Characterization of Electrically Conducting Poly-O-Methoxyaniline Zr(1V) Molybdate Cd(II) Selective Composite Cation-Exchanger," Desalination, Vol. 250, No. 2, 2010, pp. 523-529. doi:10.1016/j.desal.2008.06.033

14. S. A. Nabi, Mu. Naushad and R. Bushra, "Synthesis and Characterization of a New Organic-Inorganic Pb^{2+} Selective Composite Cation Exchanger Acrylonitrile Stannic(IV) Tungstate and Its Analytical Applications," Chemical Engineering Journal, Vol. 152, No. 1, 2009, pp. 80-87. doi:10.1016/j.cej.2009.03.033

15. T. P. Valsala, S. C. Roy, J. G. Shah, J. Gabriel, K. Raj and V. Venugopal, "Removal of Radioactive Caesium from Low Level Radioactive Waste (LLW) Streams Using Cobalt Ferrocyanide Impregnated Organic Anion Exchanger," Journal of Hazardous Material, Vol. 166, No. 2-3, 2009, pp. 1148-1153. doi:10.1016/j.jhazmat.2008.12.019

16. I. M. El-Naggar, E. S. Zakaria, I. M. Ali, M. Khalil and M. F. El-Shahat, "Studies on Synthetic Polyaniline Titanotungstate and Its Applications for Cesium Treatment," Inorganic Chemistry, 2011, in Press.

17. N. E. Topp and K. W. Pepper, "Properties of Ion Exchange Resin in Relation to Their Structure, I. Titration Curves," Journal of the Chemical Society, Vol. 690, 1949, pp. 3299-3303. doi:10.1039/jr9490003299

18. M. Qureshi, J. P. Gupta and V. Sharma, "Comparison of the Ion-Exchange Behaviour of Zirconium, Thorium, Vanadium, Uranium, Stannic and Titanium Tungstates," Talanta, Vol. 21, No. 1, 1974, pp. 102-106. doi:10.1016/0039-9140(74)80069-X

19. A. A. Khan and M. M. Alam, "Synthesis, Characterization and Analytical Applications of a New and Novel 'Organic-Inorganic' Composite Material as a Cation Exchanger and Cd(II) Ion-Selective Membrane Electrode: Polyaniline Sn(IV) Tungstoarsenate," Reactive & Functional Polymers, Vol. 55, No. 3, 2003, pp. 277-290. doi:10.1016/S1381-5148(03)00018-X

20. I. M. Ali, "Synthesis and Sorption Behavior of Semicrystalline Sodium Titanate as a New Cation Exchanger," Journal of Radioanalytical and Nuclear Chemistry, Vol. 260, No. 1, 2004, pp. 149-157. doi:10.1023/B:JRNC.0000027074.36548.29

21. M. Qureshi, J. P. Gupta and V. Sharma, "Comparison of the Ion-Exchange Behaviour of Zirconium, Thorium, Vanadium, Uranium, Stannic and Titanium Tungstates," Talanta, Vol. 21, No. 1, 1974, pp. 102-106. doi:10.1016/0039-9140(74)80069-X

22. T. Möller, A. Clearfield and R. Harjula, "Preparation of Hydrous Mixed Metal Oxides of Sb, Nb, Si, Ti and W with a Pyrochlore Structure and Exchange of Radioactive Cesium and Strontium Ions into the Materials," Microporous and Mesoporous Materials, Vol. 54, No. 1, 2002, pp. 187-199. doi:10.1016/S1387-1811(02)00320-7

23. I. M. El-Naggar and M. M. Abou-Mesalam, "Novel Inorganic Ion Exchange Materials Based on Silicates; Synthesis, Structure and Analytical Applications of Magneso- Silicate and Magnesium Alumino-Silicate Sorbents," Journal of Hazardous Material, Vol. 149, No. 3, 2007, pp. 686-692. doi:10.1016/j.jhazmat.2007.04.029

24. E. S. Zakaria, I. M. Ali and I. M. El-Naggar, "Thermodynamics and Ion Exchange Equilibria of Gd^{3+}, Eu^{3+} and Ce^{3+} Ions on H^+ Form of Titanium(IV) Antimonate," Colloids and Surfaces A, Vol. 210, No. 1, 2002, pp. 33- 40. doi:10.1016/S0927-7757(02)00216-9

25. E. S. Zakaria, I. M. Ali and H. F. Aly, "Kinetic Aspects and Swelling Changes of Magnesium and Cerium Titano-Antimonates in Aqueous and Mixed Solvents," Journal of Colloid and Interface Science, Vol. 338, No. 2, 2009, pp. 346-352. doi:10.1016/j.jcis.2009.06.031

26. I. M. El-Naggar, E. S. Zakaria, I. M. Ali, M. Khalil and M. F. El-Shahat, "Kinetic Modeling Analysis for the Removal of Cesium Ions from Aqueous Solutions Using Polyaniline Titanotungstate," Arabian Journal of Chemistry, Vol. 5, No. 1, 2010, pp. 109-119. doi:10.1016/j.arabjc.2010.09.028

27. I. M. Ali, E. S. Zakaria, M. M. Ibrahim and I. M. El-Naggar, "Synthesis, Structure, Dehydration Transformations and Ion Exchange Characteristics of Iron-Silicate with Various Si and Fe Contents as Mixed Oxides," Polyhedron, Vol. 27, No. 1, 2008, pp. 429-439. doi:10.1016/j.poly.2007.09.034

28. R. G. Dosch, N. E. Brown, H. P. Stephens and R. G. Anthony, Sandia National Laboratories Report, SAND- 92-2737C, 1992.

29. I. M. Ali, "Sorption Studies of ^{134}Cs, ^{60}Co and $^{152 + 154}$Eu on Phosphoric Acid Activated Silico-Antimonate Crystals in High Acidic Media," Chemical Engineering Journal, Vol. 155, No. 3, 2009, pp. 580-585. doi:10.1016/j.cej.2009.07.050

30. S. A. Shady, "Selectivity of Cesium from Fission Radionuclides Using Resorcinol-Formaldehyde and Zirconyl-Molybdopyrophosphate as Ion-Exchangers," Journal of Hazardous Material, Vol. 167, 2009, pp. 947-952.

31. M. G. Marageh, S. W. Husaina and A. R. Khanchi, "The Use of Clinoptilolite and Its Sodium form for Removal of Radioactive Cesium, and Strontium from Nuclear Wastewater and Pb^{2+}, Ni^{2+}, Cd^{2+}, Ba^{2+} from Municipal Wastewater," Applied Radiation and Isotopes, Vol. 50, No. 4, 1999, pp. 655-660. doi:10.1016/S0969-8043(98)00134-1

32. S. Lahiri, K. Roy, S. Bhattacharya, S. Maji and S. Basu, "Separation of ^{134}Cs and ^{152}Eu Using Inorganic Ion Exchangers, Zirconium Vanadate and Ceric Vanadate," Applied Radiation and Isotopes, Vol. 63, No. 3, 2005, pp. 293-297.doi:10.1016/j.apradiso.2005.03.007

33. S. A. I. Khan and A. A. Khan, "Synthesis, Characterization and Ion-Exchange Properties of a New and Novel 'Organic-Inorganic' Hybrid Cation-Exchanger: Nylon- 6,6, Zr(IV) Phosphate," Talanta, Vol. 71, No. 2, 2007, pp. 841-847.doi:10.1016/j.talanta.2006.05.042

34. Z. Alam, Inamuddin and S. A. Nabi, "Synthesis and Characterization of a Thermally Stable Strongly Acidic Cd(II) Ion Selective Composite Cation-Exchanger: Polyaniline Ce(IV) Molybdate," Desalination, Vol. 250, No. 2, 2010, pp. 515-522. doi:10.1016/j.desal.2008.09.008

35. A. M. Khan, S. A. Ganai and S. A. Nabi, "Synthesis of a Crystalline Organic–Inorganic Composite Exchanger, Acrylamide Stannic Silicomolybdate: Binary and Quantitative Separation of Metal Ions," Colloids and Surfaces A, Vol. 337, No. 1-3, 2009, pp. 141-145.doi:10.1016/j.colsurfa.2008.12.012

36. H. M. F. Freundlich, "Uber Die Adsorption in Losungen, Zeitschrift fur Physikalische Chemie," Leipzig, Vol. 57A, 1906, pp. 385-470.

37.   I. Langmuir, "The Constitution and Fundamental Properties of Solids and Liquids," Journal of the American Chemical Society, Vol. 38, No. 11, 1916, pp. 2221-2295.doi:10.1021/ja02268a002

38.   M. S. Bilgili, "Adsorption of 4-Chlorophenol from Aqueous Solutions by Xad-4 Resin: Isotherm, Kinetic, and Thermodynamic Analysis," Journal of Hazardous Materials, Vol. 137, No. 1, 2006, pp. 157-164. doi:10.1016/j.jhazmat.2006.01.005

39.   R. E. Treybal, "Mass Transfer Operations," McGraw-Hill, New York, 1987.

40.   B. Al-Duri, "Use of Adsorbents for the Removal of Pollutants from Wastewaters," In: G. McKay, Ed., CRC Press, 1996, p. 133.

41.   I. M. El-Naggar, G. M. Ibrahima, E. A. El-Kadya and E. A. Hegazyb, "Sorption Mechanism of Cs^+, Co^{2+} and Eu^{3+} ions onto EGIB Sorbent," Desalination, Vol. 237, No. 1, 2009, pp. 147-154. doi:10.1016/j.desal.2007.11.057

42.   D. Mohan and K. P. Singh, "Singleand Multi-Component Adsorption of Cadmium and Zinc Using Activated Carbon Derived from Bagasse—An Agricultural Waste," Water Research, Vol. 36, No. 9, 2002, pp. 2304-2318. doi:10.1016/S0043-1354(01)00447-X

43.   D. Mohan and S. Chander, "Single, Binary, and Multicomponent Sorption of Iron and Manganese on Lignite," Journal of Colloid and Interface Science, Vol. 299, No. 1, 2006, pp. 76-87. doi:10.1016/j.jcis.2006.02.010

44.   S. M. Hasany, M. M. Saeed and M. Ahmed, "Sorption and Thermodynamic Behavior of Zinc(II)-Thiocyanate Complexes onto Polyurethane Foam from Acidic Solutions," Journal of Radioanalytical and Nuclear Chemistry, Vol. 252, No. 3, 2002, pp. 477-484.doi:10.1023/A:1015890317697

45.   I. M. El-Naggar and M. M. Abou-Mesalam, "Novel Inorganic Ion Exchange Materials Based on Silicates; Synthesis, Structure and Analytical Applications of Magneso- Silicate and Magnesium Alumino-Silicate Sorbents," Journal of Hazardous Material, Vol. 149, No. 3, 2007, pp. 686-692. doi:10.1016/j.jhazmat.2007.04.029

46.   S. Netpradit, P. Thiravetyan and S. Towprayoon, "Evaluation of Metal Hydroxide Sludge for Reactive Dye Adsorption in a Fixed-

Bed Column System," Water Research, Vol. 38, No. 1, 2004, pp. 71-78. doi:10.1016/j.watres.2003.09.007

47.   I. M. Ali, "Sorption Studies of [134]Cs, ^{60}Co and [152+154]Eu on Phosphoric Acid Activated Silico-Antimonate Crystals in High Acidic Media," Chemical Engineering Journal, Vol. 155, No. 3, 2009, pp. 580-585. doi:10.1016/j.cej.2009.07.050

48.   T. J. Trantera, R. S. Herbsta, T. A. Todda, A. L. Olsona and H. B. Eldredge, "Evaluation of Ammonium Molybdophosphate-Polyacrylonitrile (AMP-PAN) as a Cesium Selective Sorbent for the Removal of ^{137}Cs from Acidic Nuclear Waste Solutions," Advances in Environmental Research, Vol. 6, No. 2, 2002, pp. 107-121. doi:10.1016/S1093-0191(00)00073-3

49.   F. Helfferich, "Ion Exchange," McGraw-Hill, New York, 1962.

50.   I. M. El-Naggar, E. I. Shabana and M. I. El-Dessouky, "Ion Exchange Behaviour of Hydrous Tin Oxide: Kinetics of Anion Exchange," Talanta, Vol. 39, No. 6, 1992, pp. 653-657. doi:10.1016/0039-9140(92)80076-P

51.   E. S. Zakaria, I. M. Ali and H. F. Aly, "Kinetic Study of the Isotopic Exchange of Na^+ and Zn^{2+} Ions on Iron and Chromium Titanates," Journal of Radioanalytical and Nuclear Chemistry, Vol. 260, No. 2, 2004, pp. 389-397.doi:10.1023/B:JRNC.0000027114.32878.d2

52.   M. D. Kaminski, L. Nunez, M. Pourfarzaneh and C. Negri, "Cesium Separation from Contaminated Milk Using Magnetic Particles Containing Crystalline Silicotitanates," Separation and Purification Technology, Vol. 21, No. 1-2, 2000, pp. 1-8. doi:10.1016/S1383-5866(99)00062-3.

Synthesis of Ti$_3$SiC$_2$-Bicarbide Based Ceramic by Electro-Thermal Explosion

Nasr-Eddine Chakri[1, 2], Zoubida Habes[2, 3],
Abdelaziz Toubal[2, 4], and Badis Bendjemil[1, 5]

[1]Physic Department, Badji-Mokhtar University, Annaba, Algeria

[2]Laboratory Environmentally Security and Alimentary (LASEA), Badji-Mokhtar University, Annaba, Algeria

[3]Chemistry Department, Badji-Mokhtar University, Annaba, Algeria

[4]Genie Chemistry Department, Badji-Mokhtar University, Annaba, Algeria

[5]University of Guelma, Guelma, Algeria

ABSTRACT

A polycrystalline dense Ti$_3$SiC$_2$ based ceramic material has been produced by several techniques. The effect of addition of TiC and SiC is also studied. The Ti$_3$SiC$_2$ material shows extraordinary electrical, thermal and mechanical properties. Furthermore, it shows a damage tolerance capability and oxidation resistance. In this work, we have synthesized Ti_3SiC_2 by electro-thermal explosion chemical reaction (ETE) with high current density (900 Amperes/a.u) followed by uniaxial pressure. The structural properties of the as-prepared materials are studied by x-ray diffraction (XRD), scanning electron microscope (SEM) and energy dispersive x-ray spectroscopy (EDX) techniques. The chemical cartography, imaging and electronic properties are investigated using Ultra-STEM and electron high energy loss resolution spectroscopy (EELS) techniques, respectively. The surface of Ti_3SiC_2 is characterized by means of X-ray photoelectron spectroscopy (XPS). High resolution C 1s, Si 2p, Ti 2p, Ti 3s core level spectra are explained in terms of crystallographic and electronic structure. Valence band spectrum is performed to confirm the validity of the theoretical calculations.

INTRODUCTION

The investigations of the ternary ceramic Ti_3SiC_2 show that it has exceptional properties [1] -[8] . Similar electrical and thermal conductivity at room temperature exceeds those of Ti metal; the former possesses excellent oxidation resistance (up to 1400°C), a resistance to thermal shock; a moderately low coefficient of thermal expansion. Ti_3SiC_2 also possess an unusual array of mechanical properties (including good compressive strength and a high Young's modulus all along with low hardness and some evidences for ductility) the ceramic is readily processed by standard tools. Also, it has been shown that large, highly oriented polycrystals of Ti_3SiC_2 undergo auto-deformation plastically at room temperature by means of shear and kink band formation. The phase was first synthesized by Jeitschko and Nowotny back on 1967 [9] . Recent interest in the phase is due to the availability of improved material, which leads to high purity polycrystalline single-phase samples of neartheoretical density. The structure has been reported [10] as being hexagonal in space group P 63/mmc (194) with a < 0.307 nm and c < 1.769 nm. It shows a great resistance against oxidation, extreme hardness, and above all, it can retain its strength to temperatures that makes the best superalloys available today usable. Up to now, no other material has shown such a combination of properties like machinability, strength, and ductility at elevated temperatures and nonsusceptibility to thermal shock. It has an excellent self-lubricating property, so it can be better than graphite for rotating electrical contacts for ac motors. It is a promising candidate for ceramic engines. More details about the technological importance of Ti$_3$SiC$_2$ can be found in reference [1] .

As far as the experiments of this compound are concerned, Jeitschko and Nowotny [9] have synthesized Ti$_3$SiC$_2$ via chemical reaction between TiH$_2$, Si, and graphite at 2000°C. Goto and Hirai synthesized Ti$_3$SiC$_2$ through the chemical-vapor deposition technique [11] . Very little is known about this material. Panczyk et al. [10] , shown that the melting point of Ti$_3$SiC$_2$ occurs at around 3000°C. Pampuch et al. [12] produced a hard Ti$_3$SiC$_2$-based material with a high young's modulus of 326 GPa through self-propagating high-temperature synthesis and ceramic processing. Barsoum and El-Raghy [1] fabricated polycrystalline bulk samples of Ti$_3$SiC$_2$ by reactively hot-pressing a blend of Ti, graphite, and SiC powders at 40 MPa and 1600°C for 4 h. Onodera et al. [13] [14] at high pressure measured the value of the bulk modulus at about 20666 MPa. The study of the electronic structure and chemical bonding gave a better insight into the important Ti$_3$SiC$_2$ materials properties. In order to study the electronic structure of Ti$_3$SiC$_2$, full-potential linear-muffin-tin-orbital methods (FPLMTO) was used [15] . The calculations were based on the local-density approximation (LDA) and the Hedin-Lundqvist [12] parameterization was used for the exchange and correlation potential. Basis functions, electron densities, and potentials were calculated without any geometrical approximation [15] . The radial basis functions within the muffin-tin spheres are linear combinations of radial wave functions and their energy derivatives which are computed at energies appropriate to their site and principal as well as orbital atomic quantum numbers, whereas outside the muffin-tin spheres the basic functions are combinations of Neuman or Hankel functions [16] [17] . For the sampling of the irreducible wedge of the Brillouin zone, the special k-point methods were used [18] . Each calculated eigenvalue was associated with a Gaussian broadening of width 20 mRy. Ti$_3$SiC$_2$ belongs to the D6h 4–P63/mmc space group of hexagonal crystalline structure [19] , in which, its primitive cell contains two formula units. The observed lattice parameters were a = 53.064Å and c = 517.650Å. C/a was optimized by calculating of the total energy for different c/a values at the equilibrium volume. The calculated value of c/a was around 5.77 and it was in a very good agreement with the experimental value of 5.76. The calculated value of the bulk modulus was 225 GPa and is in a good agreement with the experimental value [17] . If one compares it with TiC, which has a bulk modulus of 240 GPa [20] , one find that they are comparable; the density of states (DOS) for Ti_3SiC_2 and the corresponding band structure is shown. The lowest-lying states around 10 eV are originated from the non-metal C 2s states. If one compares this with the DOS of TiC [21] , there are also C 2s states which lie around 10.0 eV. Between 9.0 and 6.0 eV below the E_F, there are states which are derived from the Si 3s states. However, one can see a big gap in TiC in this energy range, due to the absence of Si. The states just below E_F are dominated by strongly hybridizing bonding states combinations of Ti 3d orbitals of eg symmetry, C and Si p-derived orbitals. The states above E_F contain antibonding Ti 3d -dominated orbitals of eg symmetry, C and Si p states. In the case of TiC, the states near E_F show the same character. E_F in TiC lies exactly in the pseudo gap, whereas in Ti_3SiC_2 it has moved outside this gap. This indicates that TiC should be harder than Ti_3SiC_2.

In summary, it is shown that the modern electronic structure theory is sufficiently developed to give an accurate description of Ti_3SiC_2. The calculated volume and structural parameters are in a very good agreement with the experimental values. The chemical bonding appears to be similar to that of TiC.

A polycrystalline material based on a Ti_3SiC_2 compound and produced by sintering the powders synthesized through the electro-thermal explosion (ETE) method has been shown to have the advantages of ceramic materials combined with some metal-type plastic properties. Reaction in ternary Ti-Si-C system under the ETE regime may produce either TiC, SiC, $TiSi_2$, Ti_5Si_3C or Ti_3SiC_2: each in a quantity depending on the starting mixture composition [22] . However, until now, attempts to produce a Ti_3SiC_2 single phase resulted at the very most in 10% TiC amount and traces of titanium-silicon intermetallic as ascertained. Polycrystalline dense Ti_3SiC_2 based ceramic has been produced by several techniques and the effect of addition of TiC and SiC is also studied [23] -[33] .

EXPERIMENTAL PROCEDURES

Polycrystalline high-purity Ti, C, and Si powders (99.99% pure) of the nominal composition 3Ti/Si/2C, was mechanically alloyed. The powder was prepared from Goodfellow SARL powder in a high-energy Fritsch planetary ball mill with a balls/mass ratio of 13/1. The rotation speed could be varied within the range 450 - 650 rpm. In order to avoid oxidation during alloying, the ball mill was filled with high-purity argon gas. The vial was opened in 10 - 12 h to assure high homogenization and repeated fractioning,. The powders were compacted into small discs (2 - 3- 13 mm) at the compacting pressure P = 6000 psi and then subjected to ETE in Argon or in vacuum at 900 A° during 55 seconds followed by uniaxial pressure.

The structural properties were determined by XRD using a Philips diffract-meter (CoKα radiation). Further structural characterizations were carried out by energy-dispersive X-ray microanalysis (EDX), scanning electron microscopy (SEM), field emission scanning electron microscopy (FESEM), and optical microscopy (OM). SEM, FESEM micrographs were taken with a JEOL JSM-840A instrument equipped with an EDX accessory to check the elemental composition of the system [33] . XPS analysis were carried out in an ESCALAB Mk.II instrument using Al Kα X-rays as the source of excitation, the gun being run at 13 kV and 20 mA. Survey spectra were obtained over a kinetic energy range of 80 - 600 eV, in steps of 0.5 eV (CAE mode) of 100 eV. Detailed data were obtained for selected ranges of photoelectron energies (i.e. Ti 2p , C 1s , O 1s and Si 2p lines) in steps of 0.2 eV and at a pass energy of 20 eV; the regions were scanned repetitively in order to improve the signal-to-noise ratio. X-ray induced AES (XAES) structures data of Si KLL were obtained from excitation with Bremsstrahlung radiation; repetitive scans were carried out at pass energy of 20 eV and for steps of 0.2 eV. The XPS data were processed by subtraction of aShirley-type background; minor charge shifting was accounted for by setting the graphitic adventitious C 1s component to 284.6 eV and by ensuring internal self-consistency of the shifted spectra. The noisiest of the spectra (e.g. Si KLL spectra) were smoothed, but not otherwise processed. The spectra from as-mounted, specimens were characterized by surface oxide and some adventitious surface carbon. Both were substantially removed by a light ion etch (ca. 6 mA min cm^{-2} dose of 3 keV) [6] .

In this work, we have produced Ti_3SiC_2 with chemical reaction by electro-thermal explosion with high currant density (900 A°) followed by uniaxial pressure. The present study was examined the structural properties of materials using XRD, SEM and EDX. The chemical cartography, imaging and electronic properties will investigated using Ultra-STEM and electron high energy loss resolution spectroscopy (EELS), respectively. The Ti_3SiC_2 is studied by means of X-ray photoelectron spectroscopy. High resolution C 1s, Si 2p , Ti 2p ,Ti 3s core level spectra are inspected in terms of crystallographic and electronic structure. Valence band spectra will be performed to confirm the validity of the theoretical calculations.

RESULTS AND DISCUSSION

The final product has been investigated and analyzed by means of scanning electron microscopy (SEM), XPS, X-ray diffraction (XRD) and EDX. The observations have proved that the elongation slabs with rounded corners of well fused Ti_3SiC_2 grains form a matrix within which some rounded TiC and less frequent angular SiC inclusions are present (Figure 1). Figure 1 shows the SEM images of the Ti_3SiC_2, bulkbicarbide base ceramic (TiC, SiC) produced via electro-thermal-explosion reaction (ETE, Figure 2) of crystalline elemental powder of Ti, C, and Si after balls milling (13/1 (balls/mass)) has been synthesized (Figure 3).

The phase composition after ETE reaction has been studied and analysed using the X-ray diffraction technique (Figure 4).

The morphology of the as-prepared materials was observed by scanning electron microscopy (SEM) as shown in the Figure 1. Further studies have been carried out by energy-dispersive X-ray microanalysis (EDX) (Figure 5) and X-ray photoelectron spectroscopy (XPS) (Figure 6).

(a)

(b)

Figure 1: SEM imaging: (a, b) representation of bulk Ti_3SiC_2 corresponding to the carbides TiC, SiC and may be intermetallic $TiSi_2$.

(a)

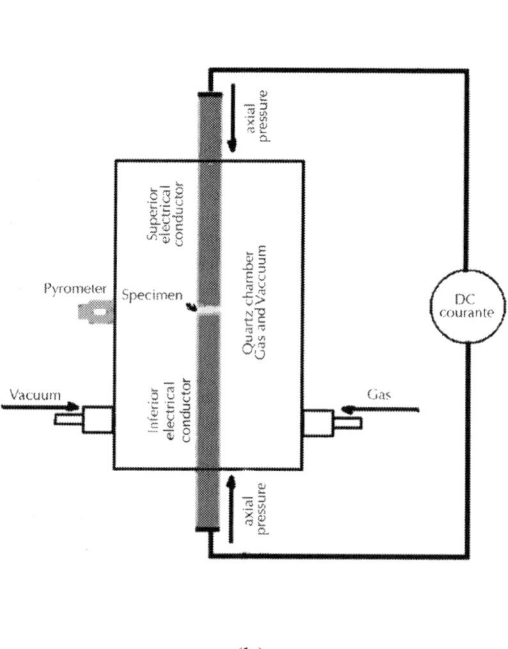

(b)

Figure 2: The electro-thermal explosion (ETE) reactor. (a) Chamber electro-thermal explosion reactor; (b) Schematic diagram of the electro-thermal explosion.

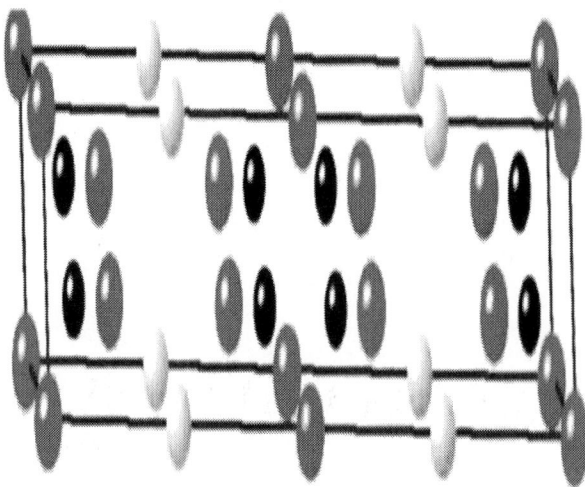

Figure 3: Crystal structure of Ti$_3$SiC$_2$, as refined from FPLO ab-initio calculations. Red, blue, and black spheres represent Ti, Si, and C atoms, respectively.

The diffraction pattern of the combustion product formed upon ETE of the starting mechanically alloyed powder of Ti, C, and Si confirms the presence of Ti$_3$SiC$_2$ bulk bicarbide base ceramic (TiC, SiC) and possibly TiSi$_2$. The morphology and microstructure of the ETE product are shown in Figure 1. The EDX data (Figure 5) also confirm the presence of the Ti$_3$SiC$_2$ bulk bicarbide base ceramic.

The formation of the Ti$_3$SiC$_2$ bulk bicarbide is a two-step process: 1) the diffusion of Si into TiC to form the Ti$_3$SiC$_2$ bulk bicarbide, and 2) the diffusion of Ti into SiC resulting in Ti$_3$SiC$_2$ bulk bicarbide. The remaining amount of TiC and SiC could be subdivided in the matrix based ceramic.

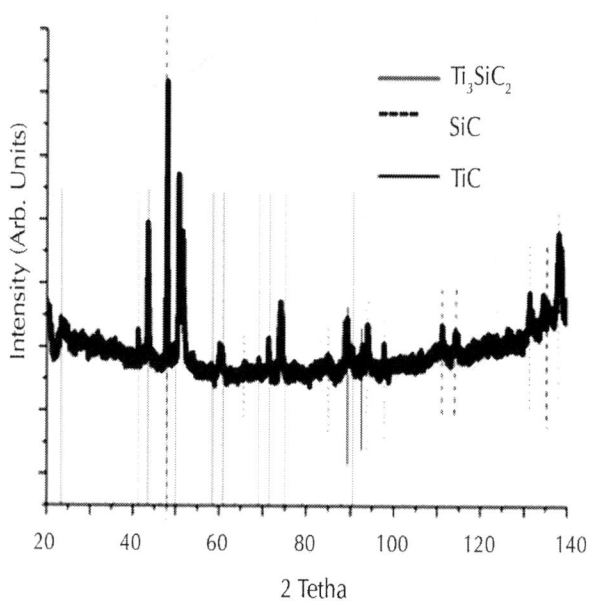

Figure 4: XRD patterns of the ETE from Ti-Si-C system product.

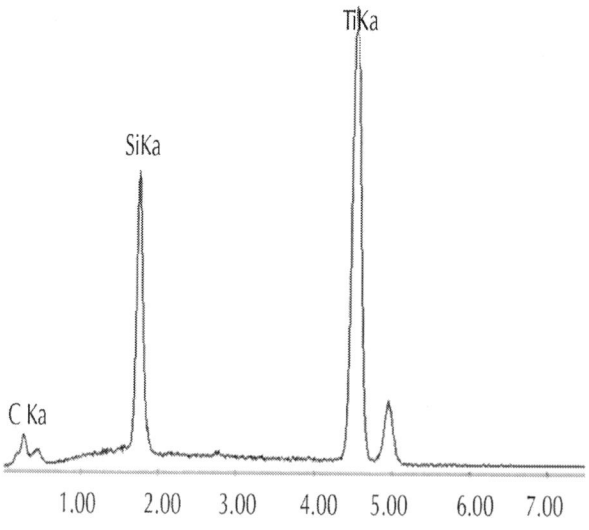

Figure 5: EDX spectra of bulk $Ti_3 SiC_2$ bicarbide based ceramic.

The Figure 6 shows the XPS data. The persistence of surface oxide is shown from the intensity of the O 1s line. The spectra of the model compounds were representative of those from the specimens that arise from optimized processing routes. Detailed spectra for the relevant specimens are shown in Figures 6(a)-(c) for C 1s, Ti 2p and Si 2p/ KLL, respectively. There were a number of significant and less intense spectra, but the latter are all well resolved. The C 1s spectra (Figure 6(a)) for Ti₃SiC₂ and (Figure 6(b)) for TiC revealed major contributions from adventitious graphitic/aromatic components at 284.6 eV, the carbide components were less intense, but well resolved at lower binding energies (BEs). The SiC spectrum, on the other hand, was dominated by a carbide peak at 283.5 eV and a small unresolved graphitic component on the high-BE shoulder of the main peak. The Ti 2p envelopes (Figure 6(a)) reflected the carbide contributions, as well as from the TiC surface that has the slightly higher BE. The detailed Si spectra (Figure 6(c)) exhibited various features. The SiC specimen resulted in envelopes (Si 2p and Si KLL) which are predominantly due to carbide, but with a small high-BE shoulder on the 2p envelope reflecting a SiOx state. The latter is matched by a resolved component in the KLL envelope at a kinetic energy (KE) of ca. 1608.5 eV. The spectra for Ti₃SiC₂ exhibited two major contributions, one being at a BE of ca. 98.5 eV with a strong Si KLL component at ca. 1617.4 eV. And the second, which is most likely due to SiOx related features, that was found at ca. 102.3 eV with a matching KLL component at ca. 1610.6 eV. The error bars on the experimental results were typically 6, 0.2 eV, with somewhat larger uncertain- ties in the data for Si KLL. The present results for SiC and TiC are in excellent agreement with those from other studies. The C 1s low-BE component at 281.06, 0.2 eV for Ti₃SiC₂ is close to the lowest C 1s value ever reported. Similarly, the Ti 2p3/2 binding energy obtained from analysis of Ti₃SiC₂ is close to that of metallic elemental Ti, i.e. 454.0 eV [6] [32] , and is less, by ca. 0.5 eV, than that for TiC. Finally, the Si 2p binding energy for Ti₃SiC₂ is below that of elemental Si, and some 2 eV is below those of the metal silicides [6] [32] and SiC.

The kinetic energy of the KLL transition is above that of elemental Si and is close to those for the metal silicides [6] [32] . The obvious qualitative conclusion is that the three constituent species are all exceptionally well screened (e.g., the Ti species appears to be as well screened as is in the metallic state), hence the low binding energies and the high Si KLL kinetic energy. The conclusion is in agreement with the observation that the ceramic is an excellent electrical conductor.

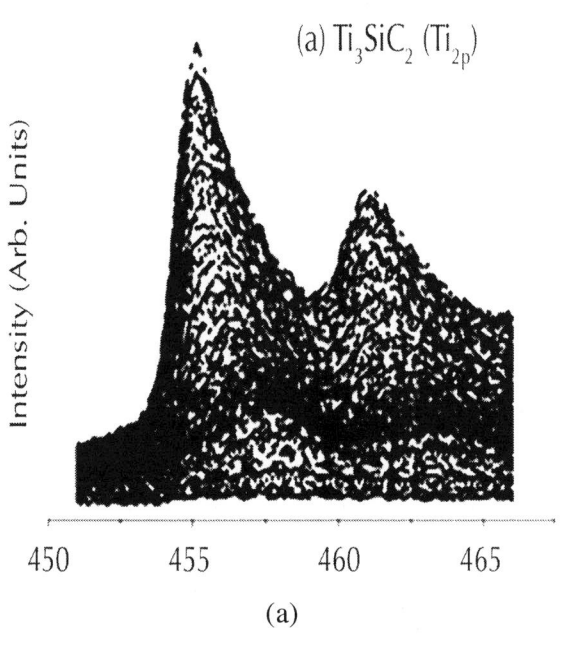

(a)

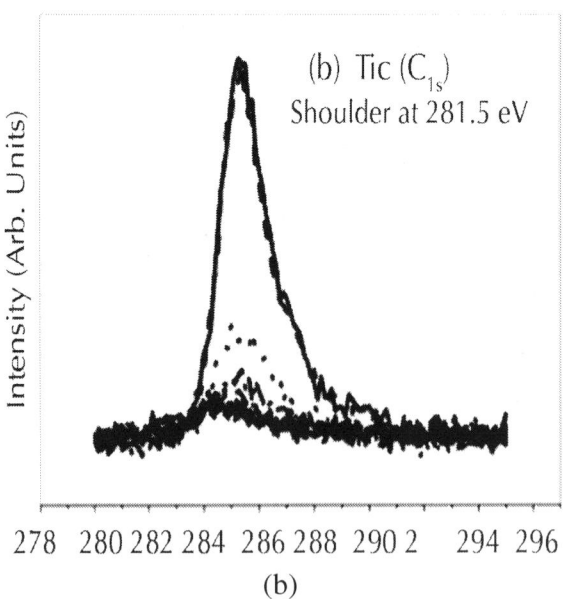

(b)

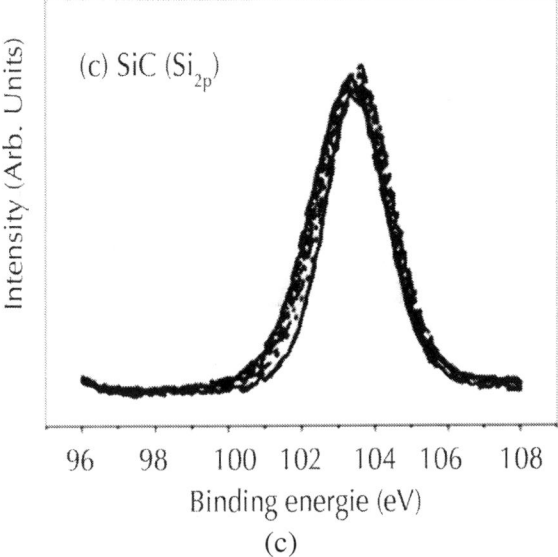

(c)

Figure 6: High resolution X-ray photoelectron spectroscopy (XPS) spectra for: (a) Ti 2p, (b) C 1s and (c) Si 2p/Si KLL. The spectra is refer to Ti$_3$SiC$_2$, TiC and SiC in order from the top to the bottom.

CONCLUSIONS

We have successfully produced that the Ti$_3$SiC$_2$ bulk bicarbide based ceramic (TiC, SiC) via ETE-method Further experiment will be focused on the synthesis of the pure Ti$_3$SiC$_2$ bulk bicarbide with ETE and SHS reactions combined ETE-Sintering and ball milling-ETE. The aim of these investigations was to contribute to understanding the properties and to synthesis a single phase of these ceramic machinable materials.

ACKNOWLEDGEMENTS

We are grateful to Dr. M. Schubert, Institute of Metallic Materials, IFW-Dresden, Germany for his help in samples analysis (SEM/EDX). I also acknowledge the head of the Spectroscopy group at the IFF-IFW-DresdenGermany, Professor Martin Knupfer for the X-ray photoelectron spectroscopy (XPS) measurements.

REFERENCES

1. Barsoum, M.W. and El-Raghy, T. (1996) Synthesis and Characterization of a Remarkable Ceramic: Ti_3SiC_2, Journal of the American Ceramic Society, 79, 1996, 1953-1956.http://dx.doi.org/10.1111/j.1151-2916.1996.tb08018.x

2. El-Raghy, T., Zavaliangos, A., Barsoum, M.W. and Kalidinidi, S. (1997) Damage Mechanisms around Hardness Indentations in Ti_3SiC_2. Journal of the American Ceramic Society, 80, 1997, 513-520. http://dx.doi.org/10.1111/j.1151-2916.1997.tb02861.x

3. Barsoum, M.W., El-Raghy, T. and Ogbuji, L. (1997) Oxidation of Ti_3SiC_2 in Air. Journal of the Electrochemical Society, 144, 1997, 2508-2516.http://dx.doi.org/10.1149/1.1837846

4. Barsoum, M.W. and El-Raghy, T. (1997) Diffusion Kinetics of the Carburization and Silicidation of Ti_3SiC_2. Journal of Materials Synthesis and Processing, 5, 1997, 197-216.

5. El-Raghy, T. and Barsoum, M.W. (1998) Diffusion Kinetics of the Carburization and Silicidation of Ti_3SiC_2. Journal of Applied Physics, 83, 1998, 112-120.http://dx.doi.org/10.1063/1.366707

6. Kisi, E.H., Crossley, J.A.A., Myhra, S. and Barsoum, M.W. (1998) Structure and Crystal Chemistry of Ti_3SiC_2. Journal of Physics and Chemistry of Solids, 59, 1437-1443.http://dx.doi.org/10.1016/S0022-3697(98)00226-1

7. Low, I.M., Lee, S.K., Lawn, B. and Barsoum, M.W. (1998) Contact Damage Accumulation in Ti_3SiC_2. Journal of the American Ceramic Society, 81, 1998, 225-231.http://dx.doi.org/10.1111/j.1151-2916.1998.tb02320.x

8. Farber, L., Barsoum, M.W., Zavaliangos, A. and Levin, I. (1998) Dislocations and Stacking Faults in Ti_3SiC_2. Journal of the American Ceramic Society, 81, 1998, 1677-1681.http://dx.doi.org/10.1111/j.1151-2916.1998.tb02532.x

9. Jeitschko, W. and Nowotny, H. (1967) Die Kristallstruktur von Ti_3SiC_2 ein Neuer Komplexcarbid-Typ. Monatshefte für Chemie/Chemical Monthly, 98, 329-337.

10. Panczyk, J., Niemyski, T., Vinogradov, L. and Sinelnikova, V. (2000) Production of Ti_3SiC_2 Material. Applied Physics Letters, 76, 1972-1976.

11. Goto, T. and Hirai, T. (1987) chemically-Vapor Deposited Ti$_3$SiC. Materials Research Bulletin, 22, 1195-1201. http://dx.doi.org/10.1016/0025-5408(87)90128-0

12. Pampuch, R., Lis, J., Piekarczyk, J. and Stobierski, L. (1993) Ti$_3$SiC$_2$-Based Materials Produced by Self-Propagating High Temperature Synthesis and Ceramic Processing. Journal of Materials Synthesis and Processing, 1, 93-100.

13. Onodera, A., Hirano, H., Yuasa, T., Guo, N.F. and Miyamoto, Y. (1999) Static compression of Ti$_3$SiC$_2$ to 61 GPa. Applied Physics Letters, 74, 3782-3796. http://dx.doi.org/10.1063/1.124178

14. Takitani, Y., Matuki, T., Li, J.-F. and Watanabe, R. (2003) Evaluation of Ti$_3$SiC$_2$ Prepared by Mechanical Alloying. Journal of the Japan Society of Powder and Powder Metallurgy, 50, 880-884.

15. Wills, J.M., Eriksson, O., Wills, J.M. and Cooper, B.R. (1987) Synthesis of Band and Model Hamiltonian Theory for Hybridizing Cerium Systems. Physical Review B, 36, 3809-3823. http://dx.doi.org/10.1103/PhysRevB.36.3809

16. Andersen, O.K. (1975) Linear Methods in Band Theory. Physical Review B, 12, 3060-3083. http://dx.doi.org/10.1103/PhysRevB.12.3060

17. Skriver, H.L. (1984) The LMTO Method. Springer, Berlin. http://dx.doi.org/10.1007/978-3-642-81844-8

18. Chadi, D.J. and Cohen, M.L. (1973) Special Points in the Brillouin Zone. Physical Review B, 8, 5747-5753. http://dx.doi.org/10.1103/PhysRevB.8.5747

19. Pearson, W.B. (1972) The Crystal Chemistry and Physics of Metals and Alloys. Wiley-Interscience, New York, 502- 518.

20. Chang, R. and Graham, L.J. (1966) Low-Temperature Elastic Properties of ZrC and TiC. Journal of Applied Physics, 37, 3778-3786. http://dx.doi.org/10.1063/1.1707923

21. Ahuja, R., Eriksson, O., Wills, J.M. and Johansson, B. (1996) Structural, Elastic, and High-Pressure Properties of Cubic TiC, TiN, and TiO. Physical Review B, 53, 3072-3087. http://dx.doi.org/10.1103/PhysRevB.53.3072

22. Lis, J., Pampuch, R. and Stobierski, L. (1992) Reactions during SHS in a Ti-Si-C System. This International Journal Encompasses Self-Propagating High-Temperature Synthesis, 1, 401-408.

23. Jin, S.Z., Liang, B.Y., Li, J.F. and Ren, L.Q. (2007) Effect of Al Addition on Phase Purity of Ti3Si(Al)C2 Synthesized by Mechanical Alloying. Journal of Materials Processing Technology, 182, 445-449. http://dx.doi.org/10.1016/j.jmatprotec.2006.09.001

24. Zhang, Z.F., Sun, Z.M., Hashimoto, H. and Abe, T. (2003) Fabrication and Microstructure Characterization of Ti_3SiC_2 Synthesized from Ti/Si/2TiC Powders Using the Pulse Discharge Sintering (PDS) Technique. Journal of the American Ceramic Society, 86, 431-436. http://dx.doi.org/10.1111/j.1151-2916.2003.tb03317.x

25. Liang, B.Y., Wang, M.Z., Sun, J.F., Li, X.P., Zhao, Y.C. and Han, X. (2009) Synthesis of Ti SiC in Air Using Mechanically Activated 3Ti/Si/2C Powder. Journal of Alloys and Compounds, 474, L18-L21. http://dx.doi.org/10.1016/j.jallcom.2008.06.147

26. Yeh, C.L. and Shen, Y.G. (2008) Effects of SiC Addition on Formation of Ti_3SiC_2 by Self-Propagating High-Temperature Synthesis. Journal of Alloys and Compounds, 461, 654-660. http://dx.doi.org/10.1016/j.jallcom.2007.07.088

27. Zakeri, M., Rahimipour, M.R. and Khanmohammadian, A. (2008) Effect of the Starting Materials on the Reaction Synthesis of Ti_3SiC_2. Ceramics International, 35, 1553-1557. http://dx.doi.org/10.1016/j.ceramint.2008.08.011

28. Liang, B.Y., Jin, S.Z. and Wang, M.Z. (2008) Low-Temperature Fabrication of High Purity Ti_3SiC_2. Journal of Alloys and Compounds, 460, 440-443. http://dx.doi.org/10.1016/j.jallcom.2007.05.074

29. Meng, F.L., Chaffron, L. and Zhou, Y.C. (2009) Synthesis of Ti_3SiC_2 by High Energy Ball Milling and Reactive Sintering from Ti, Si, and C Elements. Journal of Nuclear Materials, 386-388, 647-649.

30. Abu, M.J., Mohamed, J.J. and Ahmad, Z.A. (2012) Effect of Excess Silicon on the Formation of Ti_3SiC_2 Using Free Ti/Si/C Powders Synthesized via Arc Melting. International Scholarly Research Network, ISRN Ceramics, 2012, Article ID: 341285, 10 Pages.

31. El Saeed, M.A., Deorsola, F.A. and Rashad, R.M. (2012) Optimization of the Ti_3SiC_2 MAX Phase Synthesis. International Journal of Refractory Metals and Hard Materials, 35, 127-131. http://dx.doi.org/10.1016/j.ijrmhm.2012.05.001

32.	Briggs, D. and Beamson, G. (1992) Primary and Secondary Oxygen-Induced C1s Binding Energy Shifts in X-Ray Photoelectron Spectroscopy of Polymers. Analytical Chemistry, 64, 1729-1736. http://dx.doi.org/10.1021/ac00039a018.

7

Innovative "Green" Tribological Solutions for Clean Small Engines

Xana Fernández-Pérez[1], Amaya Igartua[1], Roman Nevshupa[2], Patricio Zabala[3], Borja Zabala[3], Rolf Luther[4], Flavia Gili[5], and Claudio Genovesio[6]

[1]Fundación Tekniker, Avda, Otaola, Eibar, Spain

[2]Instituto de Ciencias de la Construcción Eduardo Torroja (IETcc), c/ Serrano Galvache, Madrid, Spain

[3]Abamotor Energía, SL, B. Astola, Abadiano, Spain

[4]Fuchs Europe GmbH, Mannheim D, Germany

[5]CRF StradaTorino Orbassano, Italy

[6]FIRAD S.p.A, Fabbrica Italiana Ricambi Apparati Diesel, Bagnolo, Italy

INTRODUCTION

Since its invention in the last quarter of the nineteenth century and during all the twentieth century, two-stroke engines penetrated in many industrial, automotive and handheld applications where engines with high specific power, simple design, light overall weight and low cost are required. Presently, two-stroke engines are commonly used in motorcycles, scooters, chainsaw, agricultural machinery, railways grinding machines, outboard applications, etc.

Usually, the moving parts of a two-stroke motor are lubricated either by using mixture of oil with fuel or by pumping oil from a separate tank. Both designs use total-loss lubrication method, with the oil being burnt in the combustion chamber. Therefore, the lubricating oil must meet specific requirements: it must have an optimal balance of light and heavy oil components to lubricate at high temperature, it must produce no deposits (carbon sooty and other) on moving parts, and it should be ash-less. In addition, the oil should provide good protection of moving parts at high speed under deceleration of engine with the throttle closed, when the engine usually suffers from oil-starvation.

Also, two-stroke engine produce more contaminants than four-stroke engines, due to oil burning in the combustion chamber. Therefore, it is very important to reduce these contaminations to meet ecological requirements.

Most challenging issue of the European technological strategy resides in complete substitution of fossil-based fuels and lubricating oils with renewable eco-friendly and high performance materials. Esters and polyglycols were identified as alternative base oils because of their high biodegradability, low toxicity; low ash formation and absence of polymer components, in [1]. Synthetic esters are characterized by their polar structure, high wear resistance, good viscosity-temperature behaviour, miscibility with non-fossil fuels. Esther-base oils can be blended with various components like antifoam agents, oxidation inhibitors, pour-point depressants, antirust agents, detergents, anti-wear agents, friction reducers, viscosity index improvers, etc., to create environmental friendly prototype engine oils and to meet the changing environmental requirements in low sulphur fuels and other alternative fuels and their application to engine oils.

Low metal additives content and clean-burnt characteristics result in less engine fouling with much reduced ring stick and lower levels of dirt built-up on ring grooves, skirts and under crowns. Owing to the presence of polar ester groups in the molecule which have higher adhesion to metal surface, esters have much better lubricity than hydrocarbons. The performance of the ester-based lubricating oils can be further improved by selecting a proper base oil and additive package.

Another important problem is related with performance of fuel injector system when bio fuels are used. Diesel injection nozzles consist of a body (usually in Ni-Cr steel) and needle valve (High speed steel, HSS), fitted together with very strict tolerances. The design of the nozzle, i.e., the number of orifices, their diameters, positions and drilling angles depend on specific engine application. The current trend is to use multi-hole nozzles with very small holes with diameter of only 0.10 - 0.14 mm in order to improve the fuel atomization and flow pattern.

Heat treatments are applied to the body and the needle to obtain the necessary hardness both on the surface and in the core of the parts and to face the following problems:

- fatigue failures at high stress areas due to repeated pulses of very high injection pressures;
- thermal shocks.

Adequate finishing of the orifice surfaces is very important also to optimize the erosion resistance.

The usage of new diesel blends characterized by different physical and chemical properties as compared with the traditional fuels could lead to modifications both in the choice of materials, geometry and positioning of orifices or their surface finishing to ensure the correct spray pattern. This work describes the results of our recent studies aimed at solving the problems related to the introduction of new eco-friendly oils and lubricants.

NEW PROTOTYPE ENGINE OILS

Oil Characterization

Three different synthetic ester base oils have been selected to formulate three prototype engine oils with the same additive composition. These oils are different mixtures of fully saturated polyglicol-ester and mono-ester types and designated as SEMO 4, SEMO 5 and SEMO 10. Same additive package has then been added to the three bases. After comparative characterization of these prototype oils and selection of the oil with the best tribological performance (SEMO 10), a new improved formulation was developed based on the selected lubricating oil, designated SEMO 36. In addition, conventional mineral oil for two-stroke engines was used as reference oil. The additive package composition of the reference oil is different but it is ash-free as well as the other SEMO oils.

Oil viscosity was characterized according to ASTM D-445-06 standard procedure in [3], and viscosity index was determined using ASTM D-2270-04 in [4].

Deposit forming tendency of the oils was characterized by the Coker test at 250 °C during 12 h. Some physical and rheological properties of the lubricating oils are shown in Table 1. Among the prototype lubricating oils, SEMO 10 has the lowest viscosity both at 40 and 100 °C, the highest flash point and the lowest deposit forming tendency.

Unleaded petrol (E228) and bioethanol E85 (mixture of 85% of ethanol with 15% of gasoline) were selected to test miscibility of the lubricating oils with standard and alternative fuels. For this purpose two different lubricant/fuel ratios were used. Regarding to the miscibility method A (90% lubricant in fuel), SEMO 10 as well as SEMO 5 demonstrated good miscibility both with unleaded petrol and E85. Compared to this, the results for the 2% mixtures according (method B) differed. All tested lubricants proved to be perfectly miscible with EN228 fuel, whereas only SEMO 36 demonstrated to be fully miscible with E85. According to both miscibility methods the reference oil was only miscible with EN228. SEMO 36, when compared to its original prototype SEMO 10, has a much higher viscosity. Flash point for this lubricant is lower than for SEMO 10 but still higher than 200 °C.

Table 1: Properties of the engine oils

		Ref. Oil	SEMO 4	SEMO 5	SEMO 10	SEMO 36
Density, g/ml		0.877	0.915	0.917	0.935	0.999
Viscosity @ 40°C,mm^2/s		59.5	84.9	94	45.8	113.3
Viscosity @ 100°C, mm^2/s		8.6	12.5	13.2	8.0	18.3
Viscosity index		117	144	140	147	181
Flash point, °C		120	204	190	260	218
Pour Point, °C		-21	-39	-33	-39	not tested
Deposit forming *		9	4	3	9	not tested
Miscibility Method A (90% lubricant in fuel)	EN228	Good	Good	Good	Good	not tested
	E85	Poor	Poor	Good	Good	Good
Miscibility Method B (2% lubricant in fuel)	EN228	Good	Good	Good	Good	Good
	E85	Poor	Poor	Poor	Poor	Good

- *Rating on base 10

Wettability of the surface of the cylinder liner by lubricating oil is important for corrosion- and wear-protection of the piston rings and cylinder liner at the start-up when the temperature of the components is low. In this work, the wetting characteristic of the tested oils was determined using the Sessile Drop method. The resulting contact angles of the drops of various oils on the honed surface of the cylinder liner are shown in Table 2. Same method could not be used to determine wettability of the piston ring because of the small width of the ring. Therefore, the following procedure for qualitative comparison of the wettability of the piston ring by different oils in [12] was applied: 1 µl of oil was placed on the circular flat surface of the phosphate cast iron piston ring and then, after 30 s, the extension of the oil drop along this surface was measured.

Table 2: Contact angle and oil spread distance

Oil	Contact angle on honed cast iron, (°)	Spread distance of the oil on the piston ring, (mm)
SEMO 4	46.1±3.1	5.33±0.04
SEMO 5	43.4±0.2	5.39±0.05
SEMO 10	33.1±1.1	7.01±0.12
SEMO 36	50.8±0.5	3.78±0.17

The contact angle for SEMO 36 oil on the honed cast iron was the highest among all the tested lubricating oils. The contact angles of SEMO 5 and SEMO 4 were very similar one to each other and only slightly lower than for SEMO 36. SEMO 10 had the lowest contact angle and the largest drop spread for all tested oils. The behaviour of the drop spread of the tested lubricating oils over the piston ring surface is similar to that of the contact angle, bearing in mind that large contact angle values correspond to small spread distances.

Biodegradability and toxicity of the lubricating oils were examined according to the recommendations of the Organization for Economic Co-operation and Development (OECD) in [5]. Biodegradability of lubricating oils was tested using OECD 301F Manometric Respirometry Method consisting of the measurement of oxygen uptake by a stirred solution of the test substance in a mineral medium, inoculated with micro-organisms in [6]. Toxicity of the lubricating oils was studied using "Alga, Growth Inhibition Test" OECD 201 in [7] and "Daphnia Magna" 24 h Acute Immobilisation Test OECD 202 in [8]. In the "Alga, Growth Inhibition Test", selected green algae were exposed to various concentrations of the test oils over several generations under defined conditions. Results of biodegradability test are shown in Table 3. As expected, all synthetic ester base oils successfully passed the biodegradability test, while the reference mineral oil was not biodegradable according to the standard procedure OECD 301 Biodegradation of SEMO 5 and SEMO 10 exceeded 70%. In toxicity tests both with Alga and Daphnia Magna, the oils were classified as not harmful for aquatic organisms according to the standard procedures OECD 201 and 202 (see Table 4).

Table 3: Biodegradability of oils (% of biodegraded oil) in [12]

Time (days)	**SEMO** 4	**SEMO** 5	**SEMO** 10	**REF**
7	29.4	41.3	25.2	27.6
14	36.2	68.4	53.5	38.7
21	52.0	79.6	69.6	33.4
28	61.1	81.2	75.7	51.2
Ultimate	> 60%	> 60%	> 60%	< 60%

Table 4: Results of the toxicity tests in [12]

Oil	EC50/EL50 with Alga, mg/l	Classification OECD 201	EC50/EL50 with Daphnia Magna, mg/l	Classification OECD 202
SEMO 4	>100	not harmful *	>1000	not harmful *
SEMO 5	>100	not harmful *	>1000	not harmful *
SEMO 10	>100	not harmful *	>1000	not harmful *
SEMO 36	-	-	>1000	not harmful *

- *With respect to aquatic organisms.
- EC50/EL50 is that concentration of test substance which results in a 50% reduction in either growth or growth rate relative to the control.

Tribological Evaluation According To Din 51834-2

Tribological evaluation of lubricating oils was done using ball-on-disk configuration with reciprocating motion according to the standard procedure DIN 51834-2 in [9]. Ball and disk were made of 100Cr6 steel. The ball, 10 mm in diameter, performed reciprocating motion with a stroke of 1 mm and a friction frequency 50 Hz. Normal load was 50 N during short run-in period 45 s and 300 N during the test 60 min. The ball and the disk were immersed in the lubricating oil, which temperature during the test was constant and 50 °C. Friction force was measured as function of time. Friction coefficient was calculated as the ratio of the tangential force to the normal force.

After test completion, diameter of the wear scar on the ball was measured using optical microscope, and, from this data, volume wear of the ball was calculated for each lubricating oils tested.

Evolution of the friction coefficient in friction evaluation tests is shown in Figure 1. Oils with low additive content: SEMO 4, SEMO 5 and SEMO 10 showed an interval of frictional instability after the run-in period. In the instability period, which lasted from 400 up to 800 s, there are some sharp peaks indicating damage of surface and seizure, probably due to micro-welding. The reference lubricating oil had a less pronounced instability period without sharp peaks, while SEMO 36 did not present any instability. Final values of friction coefficient after 60 min and the diameters of the wear scar on the ball are shown in Table 5.

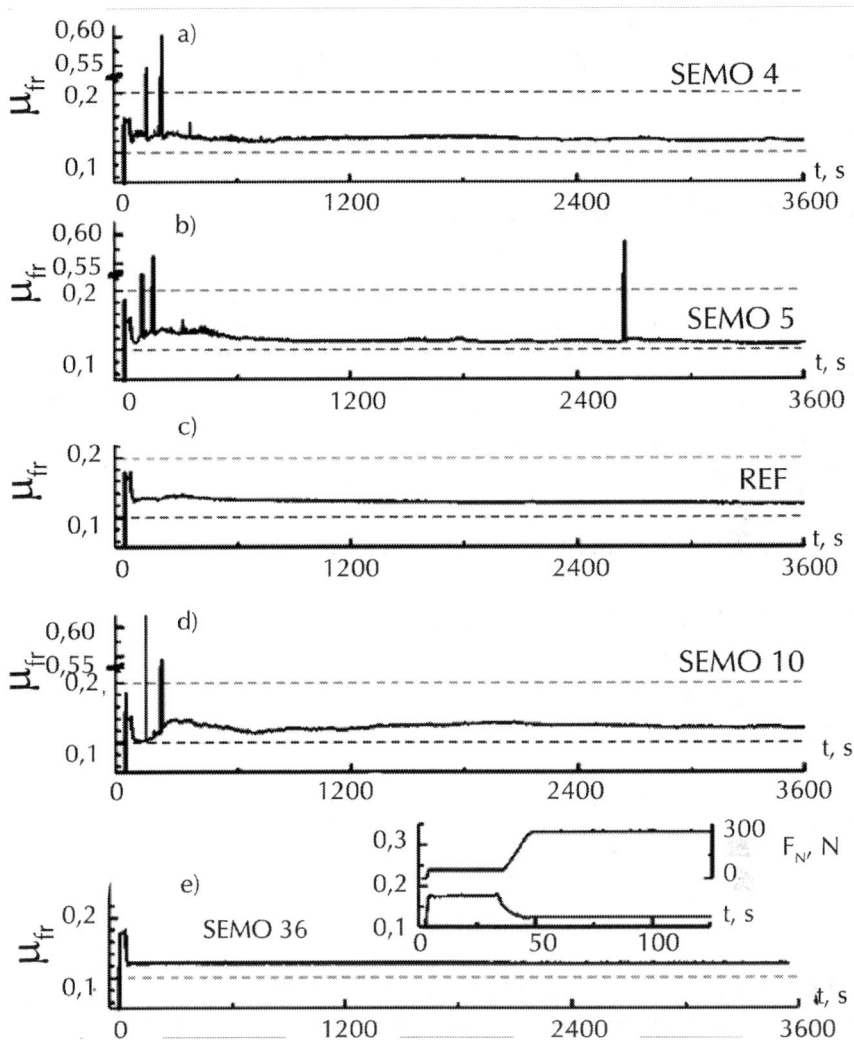

Figure 1: Evolution of friction coefficient in time during tribological evaluation tests of the following oils: a) SEMO 4, b) SEMO 5, c) reference oil, d) SEMO 10, e) SEMO 36. Inset in graph e) shows the initial part of the plot together with the curve of the normal load in [12].

Table 5: Friction coefficient and wear of the ball in tribological evaluation test of oils (DIN 51834-2) in [12]

Oil	Final μ_f^r	Diameter of wear scar, μm	Estimation of worn volume, m^{m}3
SEMO 4	0.119 ± 3.40*10^{-5}	884	6.01*10^{-3}
SEMO 5	0.111 ± 4.34*10^{-5}	885	6.02*10^{-3}
REF	0.122 ± 2.73*10^{-5}	517	6.99*10^{-4}
SEMO 10	0.125 ± 3.81*10^{-5}	929	7.32*10^{-3}
SEMO 36	0.123 ± 1.54*10^{-5}	459	4.36*10^{-4}

The volume of worn material of the ball was estimated geometrically on the basis of the diameter of the wear scar using the following equation (1):

$$V = \frac{\pi}{3} \left(R^3 - (R^2 + a^2)\sqrt{R^2 - a^2} \right)$$

(1)

where $R = 5$ mm is the radius of the ball and a is the radius of the circular wear scar.

Wear specific energy, E_w, that is, the ratio of the dissipated energy, E, during friction per unit mass of worn material m, is an important characteristic which shows the ability of a material to resist wearing. This is a complex parameter taking into account both friction, which characterizes energy supply to the material in the friction zone, and wear intensity. This parameter is considered a very useful tool to compare standard tribological evaluation and simulated tests in [10]. Wear specific energy was determined using the following equation in [12]:

$$E_w = \frac{E}{\Delta m} = \frac{v_m F_N \int_{t_i}^{t_f} \mu_{fr}(t)\,dt}{\Delta m}$$

(2)

Where v_m is mean sliding velocity obtained with a reciprocating frequency of 50 Hz and 1 mm stroke, F_N is the normal load, μ_{fr} is the friction coefficient, t_i and t_f are respectively the initial and final time points of friction test time interval.

In this study, only ball wear was determined as specified by DIN 51834-2. So, the absolute value of wear specific energy could not be determined; since wear of the disk was not measured. However, by using the ball mass loss in the denominator of eq. (2), the upper bound estimation of the wear specific energy can be determined. This upper bound can be used for qualitative comparison of anti-wear properties of the lubricating oils under constant friction conditions. These values, determined using eq. (2), are shown in Figure 2. SEMO 36 and the reference oil have much higher values of the wear specific energy, than other oils. Therefore, these lubricating oils improve contacting surfaces wear protection since much larger energy should be dissipated to produce the same wear as compared to SEMO 4, SEMO 5 and SEMO 10 lubricants.

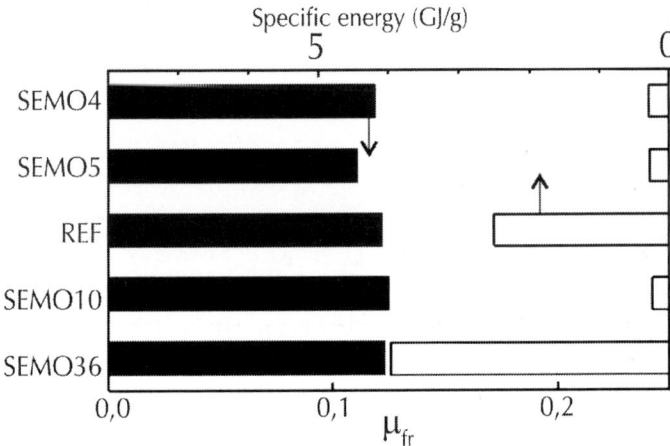

Figure 2: Friction coefficient and wear specific energy in [12].

Piston Ring/Cylinder Liner Simulation

Tribological simulation was performed using cast iron phosphated piston ring and cast iron cylinder liner using reciprocating motion configuration. The samples for the tests were cut from real engine parts (Minsel M165 two-stroke engine manufactured by Abamotor Energía) keeping original curved surfaces and surface finishing. The conformal contact between the piston ring and the cylinder counterpart was reproduced by placing a piston ring on a suitable frame, A, and fixing it by means of a clamp, B (Figure 3). Wear of the components was determined by weighting and geometry measurements. A - frame for fixing the piston ring segment, B – fixing clamp, C – base with oil bath for fixing cylinder liner sample.

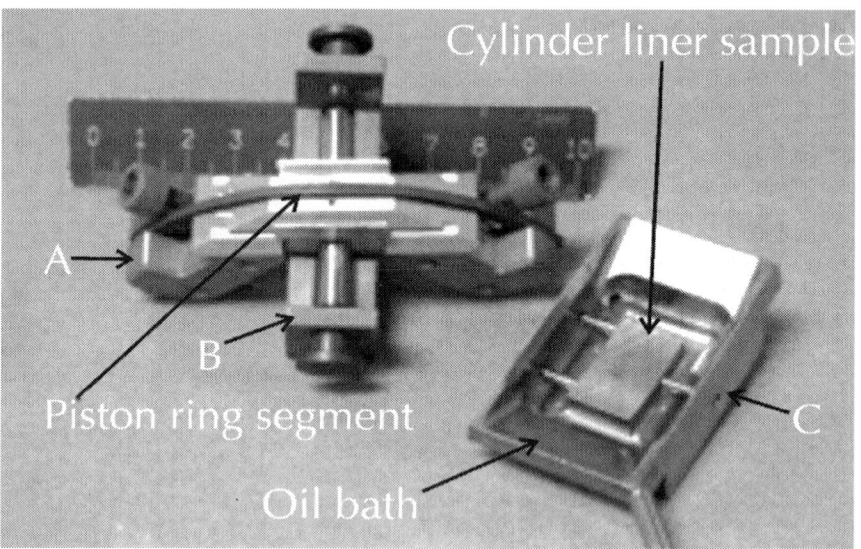

Figure 3: Experimental set-up for piston ring/cylinder liner simulation.

The piston ring segments performed a reciprocating motion with a stroke of 1 mm and a friction frequency 40 Hz. Normal load was 50 N during short run-in period 45 s and 300 N during the test 90 min. During the test, the piston ring segment and the cylinder liner sample were immersed in the oil, which temperature was constant at 200 °C.

The mass change of the piston ring segments and cylinder liner sample was determined from weighting the components before and after friction tests. Since the mass change can be due to two competitive processes: (i) wear out and (ii) deposit formation from the oil at elevated temperature, estimation of wear out by weighting can give erroneous results. Indeed, after the tests the surface colour became yellowish and remained after dissolvent cleaning indicating some sparingly soluble deposits formed on the surface due to some chemical reaction. Therefore, in addition to the determination of the mass change, worn volume was calculated from surface geometry. Surface morphology of the friction zone was studied using white light confocal microscopy at three different zones along the wear track on the cylinder liner sample. The acquired 3D surface images were 0.5 mm wide in the direction of friction and each image contained 138 cross-section profiles of the wear track yielding totally 414 profiles for each sample. Firstly, the cross-section profiles were averaged for each sample and then among different samples tested using the same lubricating oil. Worn volume of the samples of cylinder liner was calculated as a product of a mean cross-section area of the groove and the total length of the groove. The cross-section area was determined by numerical integration of the cross-section profiles and then worn mass was calculated from the worn volume using the density of cast iron.

Surface chemical composition of the friction zone of cylinder liner samples was characterized using Energy Dispersion X-Ray Spectroscopy (FDS).

Evolution of friction coefficient in time during friction between piston ring segment and a piece of the cylinder liner is shown in Figure 4. It is possible to highlight the increment of the coefficient of friction μfr for lubricants SEMO 4 and SEMO 5 overtaking the constant value reached by the lubricant of reference. In fact, for SEMO 4 and SEMO 5, friction coefficient gradually rose during the experiment (90 min) and did not stabilize. The growth behaviour was almost linear in time.

Initial friction coefficient was about 0.2 and the final one about 0.33 in both cases. SEMO 10 and SEMO 36 showed different behaviour. The initial values were 0.2 and 0.14 for SEMO 10 and SEMO 36, correspondingly.

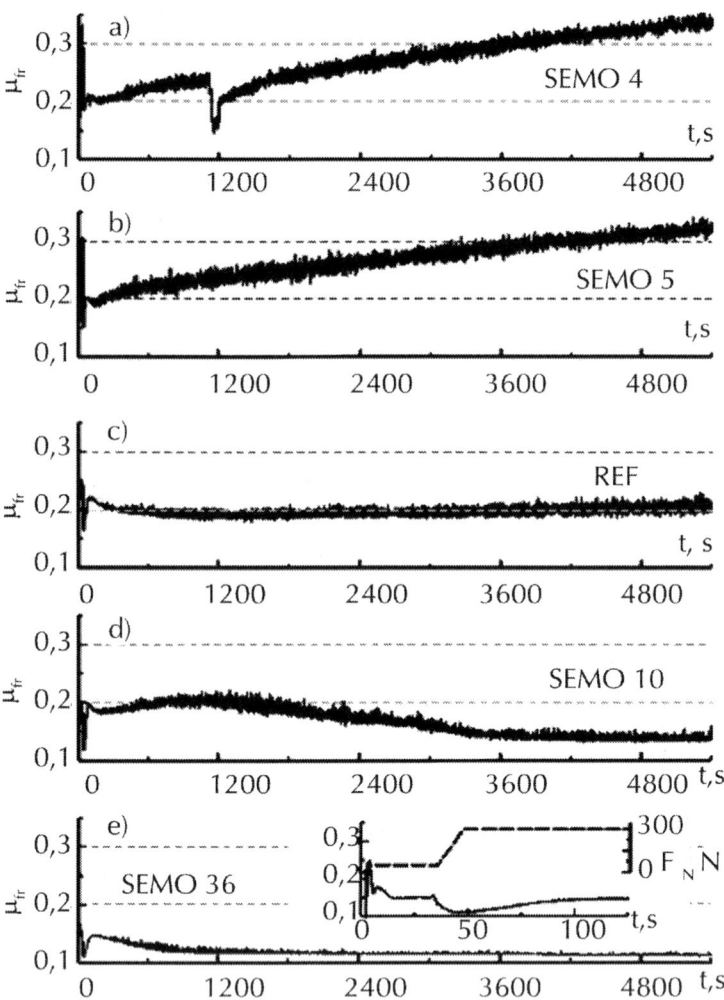

Figure 4: Evolution of friction coefficient in time during piston ring/cylinder liner simulation test. a) SEMO 4, b) SEMO 5, c) reference oil, d) SEMO 10, e) SEMO 36. Inset in the graph e) shows the initial part of the plot together with the curve of the normal load in [12].

At the beginning, after a run-in period, friction coefficient increased and reached maximum. For SEMO 36 the maximum was reached usually between 100 and 200 s from the beginning of the test, while for SEMO 10 the period of increase was longer and the maximum was reached after 700 to 1700 s from the beginning of the test. After reaching the maximum, friction coefficient decreased slowly and stabilized at 0.14 and 0.11 for SEMO 10 and SEMO 36, correspondingly. The friction coefficient of lubricant SEMO 10 showed a slow decline until reaching a constant value lower than the reference one. Friction coefficient for the improved lubricant SEMO 36 levelled out rapidly at a very low value and showed less scatter, probably due to some sort of surface deposition on the contact surfaces.

The averaged cross-section profiles of each liner sample tested are reported in Figure 5. Different scales of magnitude are used for better visualization of the mean contact surface profile. It is possible to notice very good performance of the lubricant SEMO 10 and its improvement in the lubricant SEMO 36. Samples tested using SEMO 4 and SEMO 5 had deep grooves with the maximum depth 22 to 25μm. The samples tested using the reference oil and SEMO 10 had less deep grooves with a maximum depth of 4 to 5μm. Surface of the samples tested using SEMO 36 oil had some thin scratches in the direction of friction while grooves had not been formed.

Figure 6 shows images of the friction zone of the piston ring segments after friction simulation tests with different lubricating oils. Wear and damage of the surface as function of the oil used was similar to that in the cylinder liner. In tests with SEMO 4 and SEMO 5, the material in the friction zone was heavily damaged. The wear can be classified to be of the adhesive type with intensive plastic deformation and edging. When the reference oil and SEMO 10 oil were used in the tests, the damage of the material was less pronounced than for SEMO 4 and SEMO 5, but the wear in all cases was of the adhesive type. Only small damage was observed on the piston ring segments when using SEMO 36. In this case, only summits of the circular grooves of the piston ring presented some wear and deformation. From the point of view of hydrodynamic lubrication these results may seem to be surprising, since, with the same additive composition, higher wear rate occurs for thinner oil (SEMO 10 in our case) than more viscous oils (SEMO 4 and SEMO 5). Therefore, these results lead to the following conclusions: 1) the lubrication regime should be of a boundary type and 2) surface protection against wear for SEMO 10 and SEMO 36 oils seems to be resulting from the formation of surface layer as a result of adsorption of oil components or tribochemical reactions between the oil components and the base material.

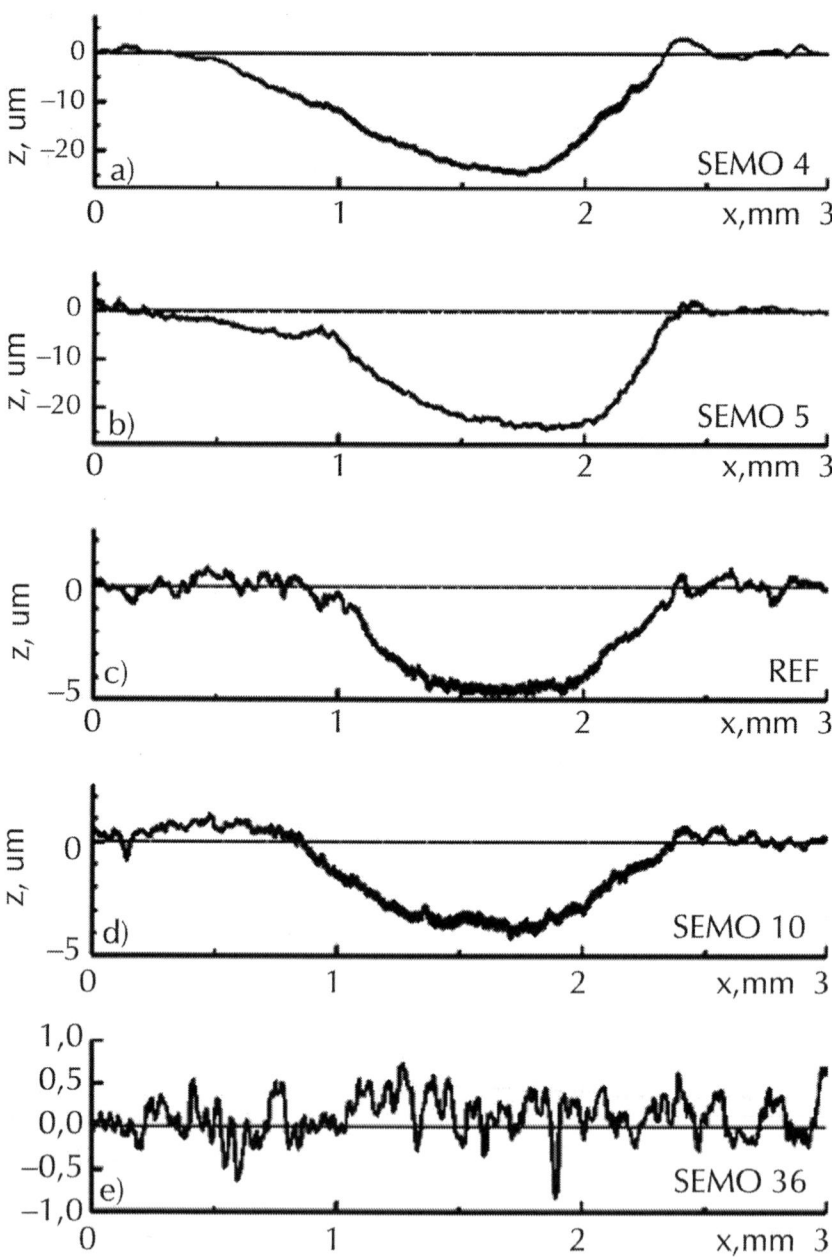

Figure 5: Average cross-section profile of the friction zone of cylinder liner samples tested using different lubricants in [12].

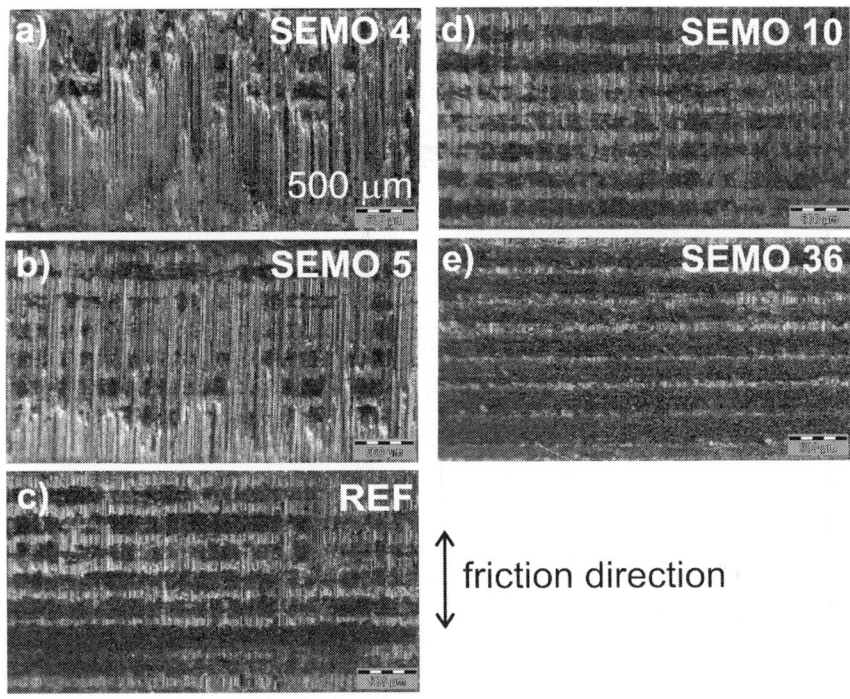

Figure 6: Optical images of the friction zone of the piston ring segments after friction simulation tests using different lubricants. The scale of each image is the same and shown by a scale bar in [12].

Results of the mass change measurements of the components are shown in Table 6. Worn mass calculated from the worn volume is plotted vs. measured mass change in Figure 7 (dots). The experimental data are fitted by linear function with two adjusted parameters: slope and intersect (dashed line in Figure 7). The solid line is a linear fit with a fixed slope 1 and adjusted intersect. Coefficients of determination for these linear regressions are 0.983 and 0.949, correspondingly, indicating statistically significant linear relationship between the mass change and worn mass determined from the geometry of the groove. Therefore, the deposit formation has not much influence on the mass change and the last can be used as a measure of the components wear out in these tests. The upper bound of the wear specific energy was determined in accordance with eq. (2), using the cylinder liner mass change in the denominator of eq. (2).

Table 6: Results of friction simulation tests in [12]

Oil	Final μ_f^r	Mass change, mg		Worn volume (cylinder.) mm^3
		cylinder	segment	
SEMO 4	0.34	-3.95	-1.41	$-1.12*10^{-2}$
SEMO 5	0.32	-3.68	-3.22	$-9.8*10^{-3}$
REF	0.20	-1.1	-0.64	$-1.74*10^{-3}$
SEMO 10	0.14	-0.94	-1.25	$-2.3*10^{-3}$
SEMO 36	0.11	-0.09	1.23	0

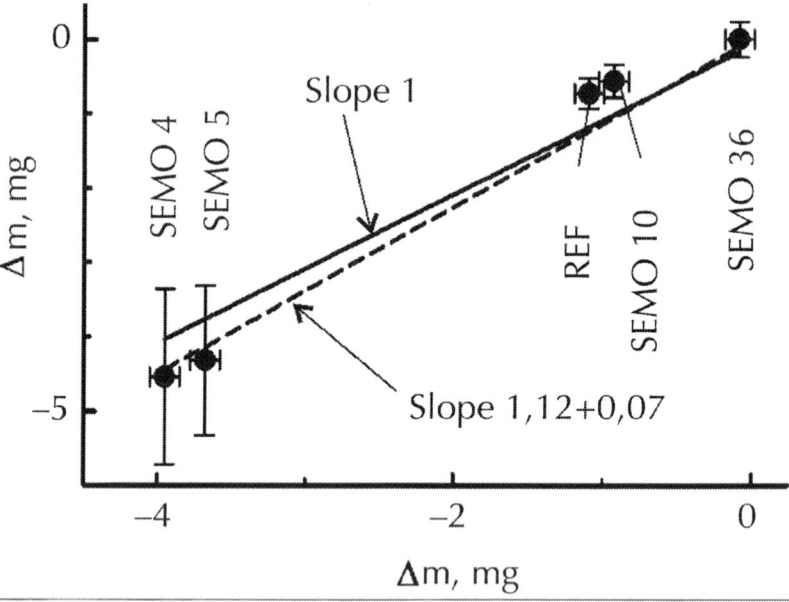

Figure 7: Mass wear determined from the geometry of the groove vs. mass change of the cylinder liner samples. The dashed line is a linear regression of experimental data with two adjusted parameters: slope and intercept. The solid line is a linear regression with a fixed slope 1 in [12].

Final friction coefficient and wear specific energy are shown in Figure 8. SEMO 36 oil showed the best antifriction and wear resistance characteristics among all tested lubricants. Friction coefficient was almost a half of that for the reference oil, while specific wear energy was 7.8 times higher than for the reference oil. In comparison with the ball-on-disk tests, wear specific energy for SEMO 36 lubricant was much lower in the tribological simulation test; however, oil temperature in these two tests was different. When the ball-on-disk evaluation tests were performed at the same temperature as in the simulation test (200 °C), the value of wear specific energy was similar to that in the simulation test: 0.14 GJ/g in the ball-on-disk at 200 °C vs. 0.18 GJ/g in the piston ring/cylinder liner simulation test. Although these values are only upper bound estimations of the real values, they are close to one another. According to the structural-energetic approach in [10], this means that the dominating wear mechanism in both cases is the same. Then, a significant decrease in the wear specific energy from 3.97 to 0.14 GJ/g with temperature increase from 50 to 200 °C implies changing in dominating wear mechanism at higher temperature. It can be stated that, under the applied experimental conditions, the chemical compositions of the base oil and the additives had greater influence on the tribological performance of the lubricants than their rheological properties.

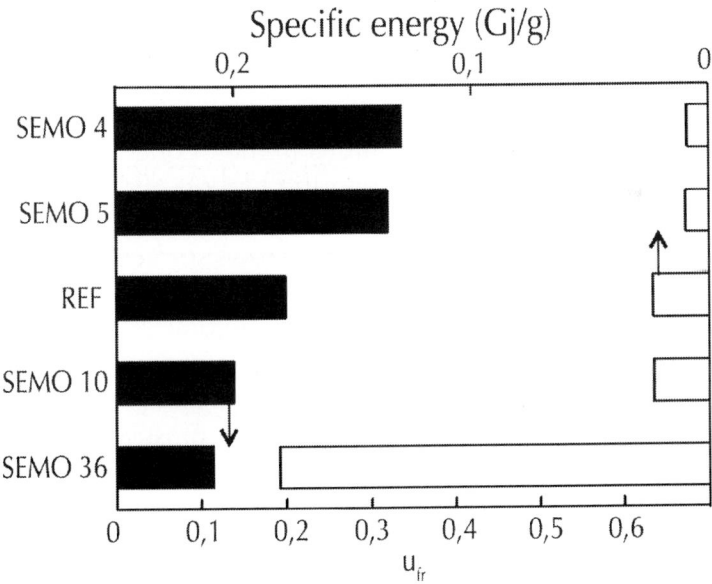

Figure 8: Friction coefficient and wear specific energy in friction simulation tests in [12].

Surface Characterization

Surface chemical composition of the friction zone of cylinder liner samples was characterized using Energy Dispersive X-Ray Spectroscopy (EDS). Table 7 shows surface chemical composition for three different surfaces:

- friction zone of the cylinder tested using SEMO 36 lubricant,
- untouched surface of the same cylinder, and
- reference cylinder not immersed neither heated in lubricating oil.

Table 7: Surface chemical composition (at.%) of the cylinder liner samples tested with SEMO 36 lubricating oil

	Fe	O	C	Si	Mn
Friction zone	61.5±1.9	11.5±0.7	23.5±3.2	2.83±0.68	0.615±0.085
Untouched zone	84.0±3.6	7.36±0.46	3.41±3.31	4.19±0.07	1.11±0.17
Reference sample	92.5±0.3	2.95±0.46	0.1	3.59±0.52	0.935±0.255

Silicon and manganese were alloying elements of the base material and did not show important variations in their concentration, whereas the most important variation was in the carbon and oxygen content. There was no significant difference for other elements since the oils had no metal-containing additives. Figure 9 shows surface concentration of four elements relative to iron. After the test, during which a cylinder was immersed in the SEMO 36 oil and heated at 200 °C, carbon and oxygen concentrations on untouched surfaces were slightly higher than on the reference sample, e.g., the sample not immersed into the oil. However, carbon and oxygen concentrations drastically increased on the surface of the friction zone, on which carbon was each forth atom. Also, in contrast to the untouched surface and the reference surface, on the surface of the friction zone, carbon concentration was higher, than the oxygen one. One can infer from these data that friction induced tribochemical reactions between oil components and base material to form surface layer enriched with carbon and oxygen. This surface layer or sliding lacquer may protect mating surfaces from adhesion and/or damage yielding lower friction and wear in [10].

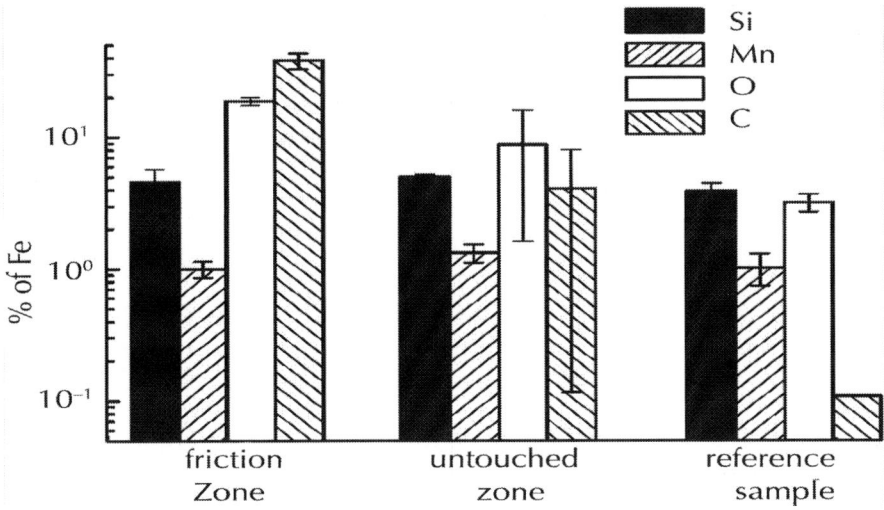

Figure 9: Surface concentration of elements relative to iron in [12].

Experimental Evaluation in Real Two-Stroke Engines (Minsel M165)

After previous simulation tribological test the performance of the oils was evaluated in real two-stroke engines (Minsel M165) with a swept volume of 158 cm³, a stroke of 54 mm, compression ratio 7,1:1, power (ISO 1585) 3.53/4.8 kW/HP, maximum torque 120 Nm and 4500 rpm rotation speed. Scuffing tests were performed using various lubricating oil – petrol mixtures in order to evaluate the lubricating performance of the lubricants under extreme load conditions. The test conditions applied are shown inTable 8, and the tested oil-fuel compositions are shown in Table 9.

Figure 10 shows the photographs of the engine components after scuffing tests, in which the reference mineral oil was used in a mixture with pure petrol and bioethanol. Increase in the bioethanol content in the fuel led to decrease in carbon soot deposition on the engine cylinder and piston. Also, when bioethanol was used, the surface was less damaged under extreme working conditions.

Table 8: Experimental conditions for scuffing tests of real two-stroke engines

Test step	Speed, rpm	Time, min	Power	
			%	HP
1	2000	5	0	0
2	4000	20	50	2.4
3	4000	20	75	3.2
4	2000	5	0	0
5	4500	90	100	full load
6	2000	5	0	0

Table 9: Oil – petrol combinations tested in a real two-stroke engine scuffing test

Fuel	Reference oil	New developed oils	
		SEMO 10	SEMO 36
Petrol	2%	2%	2%
E10	2%	-	-
E20	2%	-	-
E85	2%	-	2%

Figure 10: Macro images of two-stroke engine components after scuffing test using a mixture of mineral oil with petrol (a,d), bioethanol E10 (b,e) and bio-ethanol E20 (c,f) in [12].

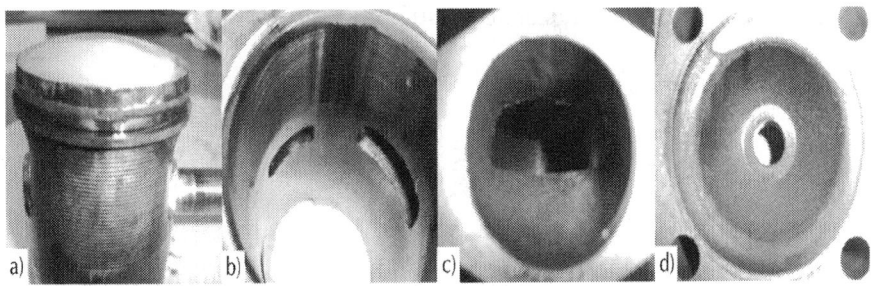

Figure 11: Macro images of two-stroke engine components after scuffing test using mixture of SEMO 10 lubricating oil with petrol: a) piston, b) cylinder, c) exhaust side, d) intake in [12].

Figures 11 show the photographs of the engine components after scuffing tests using a SEMO 10 – petrol mixture. Some seizure between compression piston ring and cylinder was observed when using a mixture of SEMO 10 with petrol. Several vertical abrasion marks were formed in the exhaust zone of the cylinder, where the temperature was higher. However, the piston and cylinder were quite clean with only some carbon soot deposits in the exhaust zone. The state of the cylinder head was quite healthy and clean in the intake zone, the carbon residues were considered normal.

Figure 12 shows the pictures of the engine components after scuffing test using SEMO 36 lubricating oil with petrol and bioethanol fuels. When using a mixture of SEMO 36 with bioethanol E85 or petrol, no scuffing or seizure was observed. Only light scratches were found on the cylinder surface, which were more pronounced when using petrol. In this case, carbon soot deposits formed intensively on the top part of the piston. The piston and cylinder were very clean, when using bioethanol.

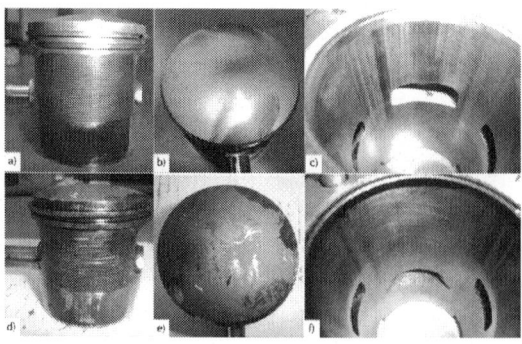

Figure 12: Macro images of two-stroke engine components after scuffing test using mixture of SEMO 36 lubricating oil with bioethanol E85 (a, b, c) and petrol (d, e, f): a), b), d), e) piston, c), f) cylinder in [12].

In addition, gaseous emissions from the engine were analyzed for various fuel-oil mixtures with different proportions of bioethanol to petrol: 20%, 30% and 85%. The gas emissions were measured using the Directive CE 2002/88, Portable, SH3 modality as reference limits. The differences in power and consumption were negligible when using bioethanol E10 and E20. When compared with the petrol, the NO_x emissions showed an increasing trend and the emissions of CO and CH diminished in tests with bioethanol and reference oil. When using E85, the reference mineral oil was not miscible, but the new developed oil SEMO 36 was totally miscible. When using bioethanol E85, a considerable reduction in engine power was observed yielding value 13% to 22% less than in the tests with petrol. At the same time fuel consumption increased slightly between 7% and 20%, and gaseous emissions were considerably reduced (see Table 10). When using SEMO 36 the reduction in NO_x emission was the most significant as compared with other gases and was probably due to the lower temperature generated.

Table 10: Emission of gases from two-stroke engine tested with different lubricating oil– petrol combinations in [12].

Oil Type and %	Power (kW)	Consumption (g/kWh)	NO_x (g/ kWh)	CH_x (g/ kWh)	CO (g/ kWh)
SH3 Limit Normative			5.36	161	603

Petrol/Ref. Oil 2%	5.46	397	1.469	139.8	333.2
E10 + 2% Ref. oil	5.44	385	1.573	124.1	314.8
E20 + 2% Ref. oil	5.5	382	2.29	128	251.5
E85 + 2% Ref. oil (not miscible)	4.8	427	2.29	109.8	43.11
E85 + 2% SEMO 36 (miscible)	4.3	478	0.689	119.5	32.93

Life Cycle

The lifecycle analysis for a 2-stroke engine fed by petrol and E85 was carried out using the model M 165 Minsel engine running in a tiller during 1000 h, which characteristics are shown in Table 11.

Table 11: Characteristics of the engine used in life-cycle analysis

Model of machine	Tiller 3002
Machine weight	90-110 kg
Engine model	M165 Minsel 2-stroke
Engine weight	12.8 kg
Engine life	1000 h
Scuffing test results	OK
Engine power	3 kW
Emissions	Directive 97/68/CE and later 2002/88/CE and 2004/26/CE

Two fuel + oil pairs named as "Cleanengine systems" were compared with the Conventional system for the same engine working in the same application. In the alternative Cleanengine system I the engine was fed by a mixture of bioethanol E20 and mineral oil. In the alternative Cleanengine system II, the engine was fed by bioethanol and newly developed advanced and biodegradable lubricating oil SEMO 36. The fuel and oil consumption for the conventional and two alternative systems is shown in Table 12.

Table 12: Parameters of the conventional and alternative systems used in the life-cycle analysis

	Conventional	Cleanengine system (I)	Cleanengine system (II)
Fuel consumption per functional unit	Petrol 900 kg	BioE20 1123 kg	BioE85 1405 kg
Oil consumption per functional unit	Mineral oil 36 kg	Mineral oil 23 kg	SEMO 36 29 kg

The Eco-indicator 99 Methodology was used for the Impact Assessment method. The components of the environmental impact are shown in Figure 13 a), while the total environmental impact is shown inFigure 13 b). Almost all components of the environmental impact as well as the total environmental impact were higher for fossil fuel. However, the climate change was more affected by the renewable system.

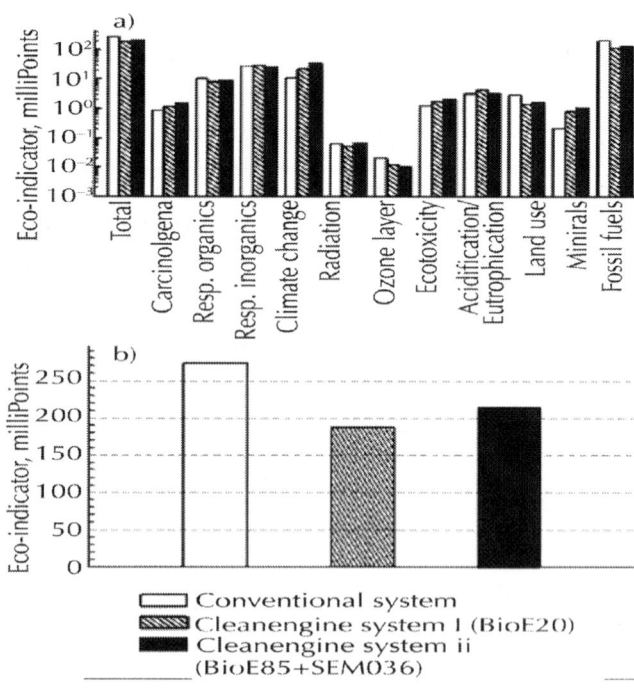

Figure 13: Results of the life-cycle and environmental impact analysis for the

conventional and two alternative systems: a) components of the environment impact, b) total environmental impact in [12].

The global environmental impact evaluated by Lifecycle Assessment tools for the Cleanengine system I and II using bioethanol was lower than for the reference system using petrol. The comparison between two alternative systems Cleanengine I and Cleanengine II showed that the last one had slightly higher environmental impact due to higher fuel and lubricant consumption that can be related to the lower calorific value of the ethanol compared to the petrol. While the reduction of the environmental impact is attributed to the reduction in emissions, the use of a biodegradable nontoxic lubricant will further reduce the environmental impact of the Cleanengine II system.

NOZZLES FOR FUTURE ENGINES

Compared to conventional liquid hydrocarbon fuels, bio-fuels exhibit considerable differences in their physical properties which significantly influence on the injector flow as well as on primary and secondary spray break-up processes. As a consequence, spray mixture formation of bio-fuels is considered to be largely different compared to conventional fuels under engine operating conditions with severe consequences on the combustion and emission characteristics. Hence, injection and combustion system optimization as well as optimization of the injector configuration (number of nozzle holes, diameter, spray targeting, etc.) for bio-fuels requires a detailed knowledge of how the fuel properties influence the injector flow and spray atomization characteristics. Optimization of the nozzles materials and design is an important task which will open new markets and enlarge the number of potential customers for eco-friendly applications.

Tribological Evaluation

Different metal-doped DLC coatings were developed by Physical Vapour Deposition method (PVD). Friction and wear tests were carried out using SRV tribometer with "cylinder-on-disc" configuration in lubricated conditions. The coatings were deposited on steel cylinders and disks. The cylinder, 15 mm in diameter, performed reciprocating motion with a stroke of 2 mm and a friction frequency 50 Hz. Normal load was 50 N during short run-in period 30 s and 200 N during the test 60 min. The cylinder and the disk were immersed in fluids, which temperature during the test was constant and 25 °C.

Both Cr- and Ti DLC coatings had good friction and wear behaviour and they could be a good alternative to improve tribological properties of the actual uncoated nozzles.

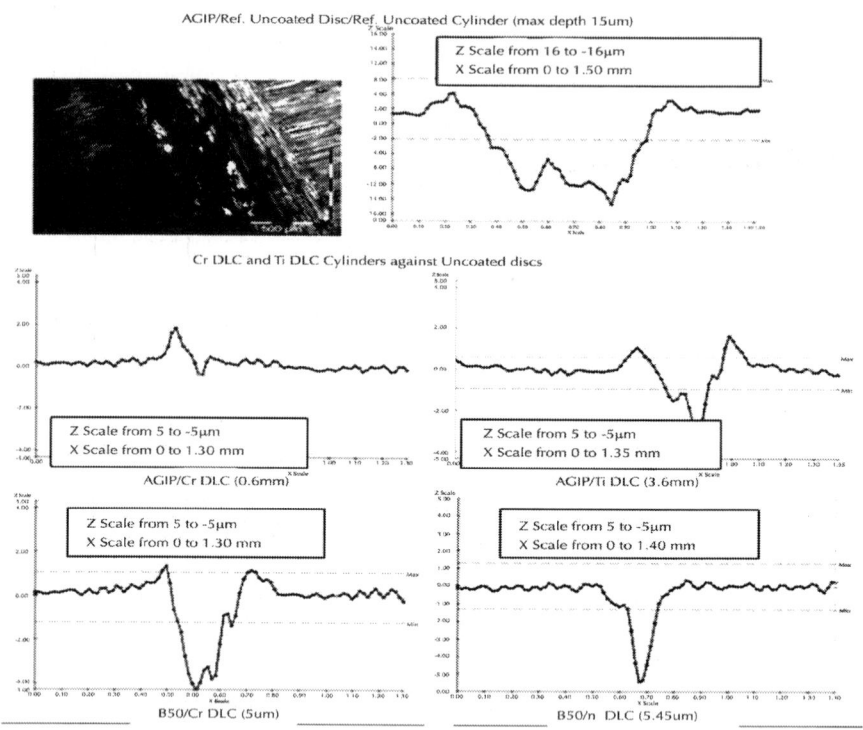

Figure 14: Average cross-section profile of the friction zone of discs samples (uncoated reference, Cr and Ti DLC) tested using different fuels, AGIP and B50. Different scales of magnitude are used for better visualization of the mean contact surface profile.

Surface morphology of the friction zone was studied using white light confocal microscopy. The averaged cross-section profiles for each sample tested are shown in Figure 14. It is possible to notice very good performance of the coatings, them had deeper grooves with a maximum depth of 5.45 µm. Cr DLC tested against AGIP fuel had better performance than Ti DLC. Two different scales of magnitude, Z, are used for better visualization of the mean contact surface profile. From 5 to -5μm for coated discs (Cr DLC and Ti DLC lubricated with AGIP ref and B50) and from 16 to -16μm for uncoated ref samples.

Corrosion Characterization

Corrosion resistance of different materials and coatings used for nozzles fabrication (Cr and Ti DLC) was characterized using electrochemical impedance spectroscopy and potentiodynamic polarization techniques in order to determine the kinetics parameters and the corrosion mechanisms of these materials in NaCl 0.5M or K_2SO_4 0.2M in [12].

Base nozzles material, uncoated steel X82WMo, was also characterized under corrosion conditions and compared with DLC coated samples of the same material. The electrolyte used in these tests was K_2SO_4 0.2M. Cr DLC coating offered excellent corrosion protection. The coating did not exhibit any pores or defects, protecting effectively the substrate during immersion.

Open-circuit potential (OCP) was measured during 2200 s in order to analyze the samples tendency with the exposure time. After that, an electrochemical impedance spectroscopy was performed in a frequencies range from 10 k to 10 mHz. Once impedance measurements finished, a potentiodynamic potential swept was applied from OCP-0.2 V to OCP+0.6 V at a scan rate of 0.5 mV/s.

Coated nozzles had more positive potential than the reference ones. For all surfaces, OCP was stable after first 2200 s of immersion. The difference between three nozzles regarding impedance results was very notable. Cr and Ti DLC coated samples had a semicircle Nyquist diagrams implying that the electrolyte did not reach the substrate during the immersion in the dissolution. The coating acted as an effective protective barrier. Uncoated nozzle had lower corrosion resistance. Two time constants could be clearly distinguished from two maxima in the Nyquist plots.

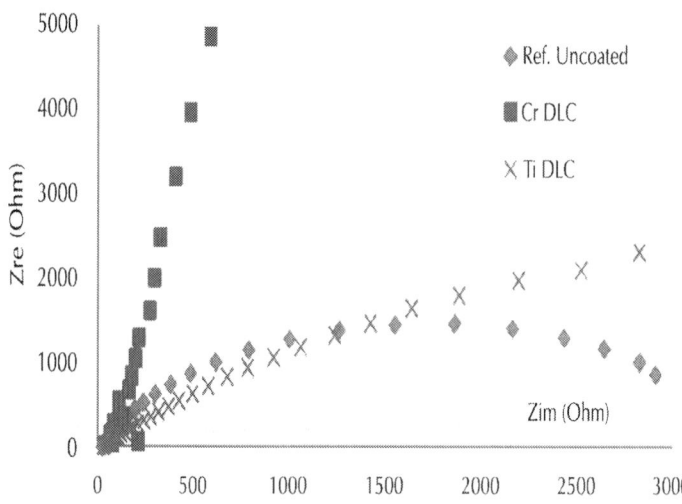

Figure 15: Nyquist diagrams. Impedance data of coated and uncoated nozzle in K_2SO_4.

Table 13 shows the parameters obtained from equivalent circuit simulation of the experimental data and Figure 16 shows the equivalent circuits used in the simulation process.

Table 13: Equivalent circuit parameters of coated and uncoated nozzles

Samples (Nozzle appl.)	OCP (V)	R1 (kΩ/ cm²)	CPE1 (µF/ cm²)	R2 (kΩ/ cm²)	CPE2 (µF/ cm²)	(µΩ−1· 1/2; 1/2)
Nozzle Uncoated	-0.556	0.035	Y_o=58.45 n=0.82	1.10	Y_o=140.7 n=0.94	
Nozzle Cr DLC	-0.022	12780	Y_o=0.75 n=0.94	-	-	-
Nozzle Ti-DLC	-0.354	10.38	Y_o=0.34 n=0.639	-	-	Y_o=0.02 B=3.23

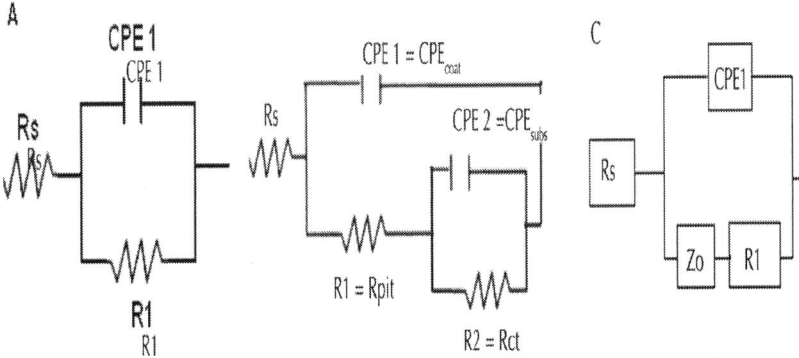

Figure 16: Equivalent circuits used for the experimental data simulation. Circuit A) for Nozzle Cr DLC; circuit B) for uncoated nozzle and circuit C) for Ti DLC coating.

Polarization curves for the coated nozzle are shown in Figure 17.

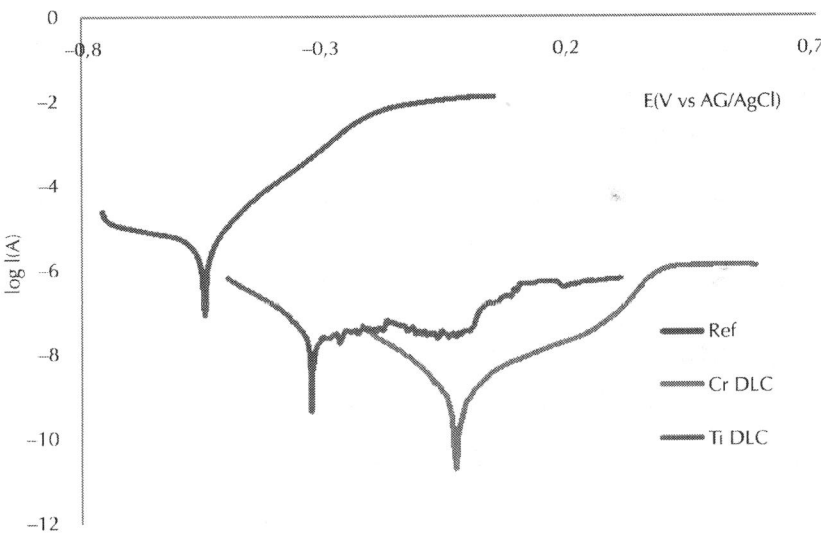

Figure 17: Polarization curves on coated and uncoated nozzles immersed in K_2SO_4

Cr DLC coating had passive behaviour and low corrosion current of the order of $10^{-9}A$ for potentials near to OCP. Coating Ti DLC also had passive behaviour in a wide zone of the anodic branch. Cr DLC and Ti DLC notably improved substrate corrosion behaviour reducing its corrosion current by several orders of magnitude (see table 14).

Table 14: Corrosion current of coated and uncoated nozzles calculated using Tafel approach

Samples	E_{corr} (V)	I_{corr} ($\mu A/cm^2$)
Nozzle Ref. Uncoated	-0.533	18.5
Nozzle Cr DLC	-0.039	0.003
Nozzle Ti DLC	-0.331	0.18

Experimental Evaluation in Real Four-Stroke Engines (Minsel M430)

The injectors were tested in the Minsel M-430 engine manufactured by Abamotor Energía, SL. The parameters of the engine and test conditions are shown in Table 15.

Table 15: Characteristics of the engine used in engine tests to evaluate the different alternative nozzles (Cr DLC and Ti DLC)

Bore	85 mm
Stroke	75 mm
Displacement	426 c.c.
Compresion ratio	19,3: 1
Power NB	8,4 / 8,7 Cv
Rpm	3000
Torque	23Nm / 2000 RPM
Dry weight	45 Kg

During the test the engine worked for 50 hours at full load (3000 rpm). Biodiesel B30 was used as a fuel, which was a mixture of FAME (100% Biodiesel) with diesel B at a rate of 30%.

Figure 18: The engine on test bench and the tested nozzles installed on the engine.

Nozzle Characterization after Test in the Engine

Scanning electron microscope (SEM) and energy dispersion X-Ray spectroscopy (EDS) were used for characterization of the nozzles geometry after the engine tests. Cr DLC coating had better behaviour than Ti DLC.

The microanalysis showed that for the all coatings the deposited layer on the needle persisted after the test, with the exception of the tip where the Ti DLC layer has been detached

Additionally, the spray holes geometries of the nozzle body were analysed after endurance test with two different fluids: reference standard fuel and 30% biodiesel.

Figure 19 shows the scanning electron microscope images (EDS) of the nozzle body tip before the engine test (real part and its corresponding silicon model for orifices internal characteristics analysis), whereas Figures 20 and 21 show the nozzles after the tests with standard diesel fuel and B30 fuel, correspondingly. Though large quantities of carbonaceous deposits could be observed on free surfaces for both fuels, no deposits were found on internal surface of spray holes.

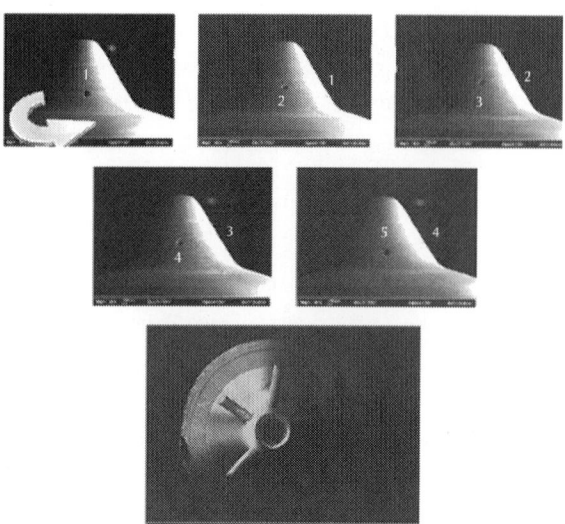

Figure 19: Nozzle body tip and silicon model with labelled holes.

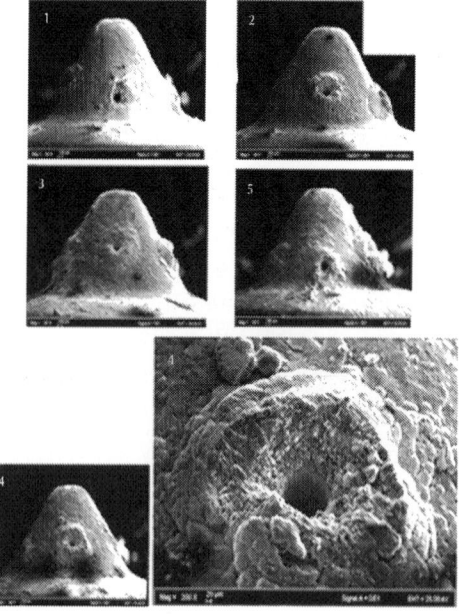

Figure 20: Images of the nozzle after endurance test with standard diesel fuel.

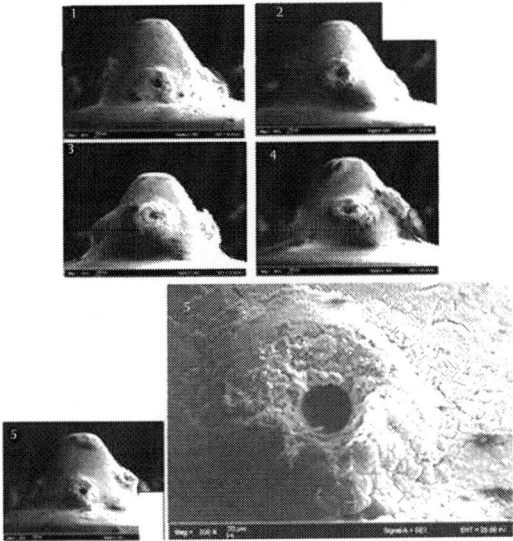

Figure 21: Images of the nozzle after endurance test with B30 fuel.

Finally, nozzle deposits were analyzed by Thermal Gravimetric Analysis (TGA), which showed no big difference in deposits composition for the nozzles operated with standard diesel and B30 blend.

CONCLUSIONS

Fully formulated prototype lubricants based on synthetic esters had low toxicity for aqueous organisms (algae and Daphnia Magna) and high biodegradability evaluated by the Manometric Respirometry Method.

Among three developed prototype lubricating oils, SEMO 10 had the best tribological performance which was comparable with that of the reference mineral oil. Further improvement of the tribological properties of this lubricating oil was achieved by additive re-formulation. The developed lubricant, SEMO 36, exceeded the reference mineral oil in tribological performance.

Our findings indicated that, in addition to the rheological properties of the lubricating oil, deposit build-up was an important factor controlling the tribological performance of the oil both in simulation experiments and real two-stroke engines. Two kinds of deposits: carbon soot and transparent sliding lacquer were observed on the engine components after tests. Build-up of a transparent sliding lacquer was especially important in the case of SEMO 36 oil and it was related with considerable reduction both in wear rate and friction coefficient. For SEMO 36, surface chemical analysis of the friction zone showed important changes in surface chemical composition, which was especially marked by increase in carbon and oxygen content. It is evident that formation of the sliding deposits stemmed from tribochemical reactions between the oil components and base material (cast iron and steel). The chemical state of carbon and oxygen atoms on the surface of friction zone should be further investigated for better understanding of these mechanisms.

Tests in real two-stroke engines were performed using mixtures of the developed lubricant with petrol or bioethanol. In both cases, no seizure between piston ring and cylinder liner was observed. When using bioethanol, the engine components were clean without important carbon soot deposits.

Engine power slightly decreased and fuel consumption slightly increased - on a volumetric basis when bioethanol E85 was blended with the newly developed lubricating oil SEMO 36. However, these results might be related with lower calorific value of ethanol as compared with petrol. Besides, the new lubricating oil improved scuffing resistance in combination with miscible lubricants and significantly reduced the environmental impact. In addition to low toxicity and high biodegradability, emissions of CO, NO_x and hydrocarbons from engines lubricated with the newly developed lubricants were lower than with traditional mineral oil and much below the limits established for portable applications.

Concluding, a new generation of lubricating oils for two-stroke engines have been developed combining low friction, good protection against wear and scuffing, no ash residue, low carbon soot or other deposit formation. These lubricating oils are compatible with bioethanol E85.

Application of Cr DLC coating on injection nozzles significantly increased the corrosion resistance and improved behaviour in engine test.

Though Ti-DLC coating also improved substrate corrosion resistance, its performance in engine test was worse than for Cr DLC coating.

Deposit chemical composition and the nozzle performance did not significantly vary in endurance tests when standard diesel was substituted by B30 blend.

ACKNOWLEDGEMENTS

The authors acknowledge financial support of the European Commission, the project CleanEngine "Advanced technologies for highly efficient Clean Engines working with alternative fuels and lubes" under contract TST5-CT2006-031241, and the Spanish Minister of Science and Innovation, for co-financing the project under contracts ENE 2008-00652-E/ALT "Tecnologías Avanzadas para motores limpios altamente eficientes, trabajando con combustibles y lubricantes renovables", RYC-2009-0412 and BIA2011-25653. Also, the authors acknowledge support received from other partners who participated in the projects: ARIZONA Chemical, OBR, GUASCOR Power, BAM, AVL and INSTITUTO MOTORI.

We appreciate the useful help of Olatz Areitioaurtena and Raquel Bayón on performing Biodegradability, toxicity and corrosion characterizations and tests.

REFERENCES

1. Igartua, A; Barriga, J; Aranzabe, A; (2005) *Biodegradable Lubricants*. Virtual Tribology Institute Edition, ISBN 83-70204-418-X.

2. Woydt, M; Skopp, A; (2005) *Ash-free and bionotox engine oils*. In: Biodegradable lubricants, eds. A. Igartua, J. Barriga, A. Aranzabe, Radom: Virtual Tribology Institute, Institute of Terotechnology; p. IV.6-IV.9.

3. ASTM D-445-06: Standard Test Method for Kinematic Viscosity of Transparent and Opaque Liquids (and Calculation of Dynamic Viscosity).

4. ASTM D-2270-04: Standard Practice for Calculating Viscosity Index from Kinematic Viscosity at 40 and 100°C.

5. OECD guidelines for testing of chemicals: Section 3; 2003. 12 p.

6.   OECD 301F: Manometric Respirometry Test. OECD guidelines for testing of chemicals; 1992.

7.   OECD 201: Alga, Growth Inhibition Test. OECD guidelines for testing of chemicals; 2006.

8.   OECD 202: Daphnia sp. Acute Immobilisation Test. OECD guidelines for testing of chemicals; 2004.

9.   DIN 51834-2. Tribological test in the translatory oscillation apparatus. Part 2: Standard Test Method for Measuring the Friction and Wear Properties of EP Lubricating Oils Using the SRV Test Machine; 2004.

10.   Kostetsky, BI; (1992) *The structural-energetic concept in the theory of friction and wear* (synergism and self-organization). Wear; 159:1-15.

11.   Nevshupa, RA ; (2009) The *role of athermal mechanisms in the activation of tribodesorption and triboluminisence in miniature and lightly loaded friction units*. Journal of Friction and Wear; 30:118-126.

12.   Igartua, A; Nevshupa, R; Fernández-Pérez, X; Conte, M; Zabala, R; Bernaola, J; Zabala, P; Luther, R; Rausch, R; (2011) *Alternative Eco-Friendly Lubes for Clean Two-Stroke Engines*. Tribology International, 44, 727-736.

13.   Martínez, L; Nevshupa, R; Álvarez, L; Huttel, Y; Méndez, J; Román, E; Mozas, E; Valdés, JR ; Jimenez, MA; Gachon, Y; Heau, C; Faverjon, F; (2009)*Application of diamond-like carbon coatings to elastomers frictional surfaces*.Tribology International, v. 42, pp. 584-590.

14.   Bayón, R; Nevshupa, R; Zubizarreta, C; Ruiz de Gopegui, U; Barriga, J; Igartua, A; (2010) *Characterisation of tribocorrosion behaviour of multilayer PVD coatings*. Analytical and Bioanalytical Chemistry.V. 396. P. 2855-2862.

15.   Bayón, R; Zubizarreta, C; Nevshupa, R; Rodriguez, JC; Fernández-Pérez, X; Ruiz de Gopegui, U; Igartua, A; (2011) *Rolling-sliding, scuffing and tribocorrosion behaviour of PVD multilayer coatings for gears application*. Industrial Lubrication and Tribology. V. 63/1. P. 17–26.

16.   Alajbegovic, A; Meister, G; Greif, D; Basara, B; (2001) *Three Phase Cavitating Flows in High Pressure Swirl Injectors*. 4th Int.

Conf. on Multiphase Flow – ICMF'01, May 27 – June 1, 2001, New Orleans, Louisiana, U.S.A.

17. Von Berg, E; Alajbegovic, A; Tatschl R; Krüger, C; Michels, U; (2001) *Multiphase Modeling of Diesel Sprays with the Eulerian/ Eulerian Approach* (DaimlerChrysler AG), ILASS-Europe 2001, Sept. 2-6, 2001, Zürich, Switzerland

18. Von Berg, E; Alajbegovic, A; Greif, D; Poredos, A; Tatschl, R; Winklhofer, E (2002); *Break-up Model for Diesel Jets based on Locally Resolved Flow Field in the Injection Hole*, ILASS-Europe 2002, Sept. 9-11, 2002, Zaragoza, Spain

Risk Analysis:
Casing-While-Drilling (CwD)
and Modeling Approach

Francisco Sánchez[a] and Mansoor H. Al-Harthy[b]

[a]Well Engineering Department, Petroleum Development of Oman (PDO), Muscat, Muscat, Oman

[b]Petroleum and Chemical Engineering Department, Sultan Qaboos University, Al-Khod, Muscat, Oman

ABSTRACT

In today's volatile economy and uncertain drilling environment, managers are encouraged to reduce well cost and time and have implemented Casing-while-Drilling (CwD) to improve operational excellence. Risk analysis is another valid tool that can be used to improve drilling operations. This paper discusses CwD as a new technology and how its benefits can be strengthened by including risk analysis as a complementary technique. A modeling approach is presented to demonstrate how risk analysis can be applied to CwD programs and to discuss the main concerns a well planner must address to achieve a successful drilling program. The integration of both CwD and risk analysis will add value to the overall excellence of the well operation. Very little work has been done on this integration, and the hope is that this approach will be a standard practice in the future.

INTRODUCTION

With today's volatile oil prices, the need to reduce the time and cost of drilling a well is becoming a major issue for drilling managers in oil and gas companies. The need to insure that costs stay within a reasonable margin cannot be ignored. Drilling managers are continuously reminded to reduce well time and cost to stay competitive in the oil market. One of the technologies that has been proven to reduce well cost and time is Casing-while-Drilling (CwD). This technology has been widely used in America and Europe but less widely in Gulf countries. This paper has two main objectives: first, to demonstrate the benefits of using Casing-while-Drilling technology instead of conventional drilling methods; and second, to integrate the risk analysis approach into CwD well cost and time forecasting. These complementary techniques, risk analysis and CwD, will allow well planners to better estimate well cost and time. Risk analysis is a tool used to capture uncertainties in input variables. This approach is more pragmatic than traditional deterministic AFE (authorization for expenditure) estimates.

This paper will focus on CwD as a new technology and why it can be a superior approach, in terms of adding value to drilling projects. Furthermore, it will introduce the risk analysis approach and why it should be used to add value to drilling operations. Finally, a model of CwD incorporating risk analysis is presented.

CASING-WHILE-DRILLING TECHNOLOGY

Casing-while-Drilling (CwD) is the technique of drilling (or reaming) with the casing connected to a special bottom-hole-assembly (BHA) instead of using a drill-pipe with a conventional set of drill-collars and a heavy drill-pipe assembly. Depending upon the CwD application, the drill-bit could be conventional or drillable, and the casing connection may be either standard API-threaded or a special modified-thread premium. Experience has demonstrated that a drilling rig's top-drive system (TDS) is the appropriate tool to transmit the rotation to the casing string while drilling, instead of using a conventional kelly-joint. CwD is a contemporary method of drilling wells, in which all drilling parameters must be rigorously controlled and closely monitored by qualified personnel, and these constraints are effortlessly met by using TDS.

Over the last decade, CwD technology has proven to be an efficient method of saving time and cost by mitigating the non-productive time (NPT) associated with several drilling problems. CwD has been successfully employed in thousands of wellbore sections worldwide (Kardos, 2008), regardless of the geometry of the wellbore, mostly in straight holes. Wellbore stability problems in horizontal or highly deviated sections have also been successfully reduced by directional CwD technology (Avery et al., 2009 and Borland et al., 2006).

Because of the effectiveness of this technique in areas that present very low probabilities of success with conventional drilling methods, more companies are now drilling with casing across troublesome, sloughing-type formations as the last available option for success. In fact, in some projects, the practice of reaching the setting depth with the casing and then cementing it into place has been highlighted by conspicuously neglecting any economic objective. In others words, CwD may increase the gross drilling cost by requiring suitable pipe drive mechanisms and special drillable bits, as well as fit for purpose casing connections, materials, and accessories. However, the overall benefit of CwD lies in reducing the drilling and tripping times, mitigating the risk of drilling problems, and consequently, decreasing the final cost of successfully delivered projects.

It has also been proven that CwD allows engineering groups to optimize well designs by applying the slim-hole concept, saving cost (30% in conjunction with other drilling optimization techniques) while developing mature hydrocarbon fields (Gordon et al., 2005). Furthermore, some engineering groups are implementing CwD not only to manage drilling across troublesome formations but also to enhance overall drilling performance, wellbore integrity, and HSE

(Health, Safety, and Environment) statistics by minimizing the risk of handling extremely heavy tubular (Hartsema et al., 2007).

Some experts agree that CwD will be used more often to enhance drilling performance in several areas by substituting for the traditional drill stem, and others are even more optimistic about this technology in believing that CwD is the future (Gupta and Banerjee, 2007). However, many negative aspects of CwD have been identified by worldwide users. Restricted weight on the bit, controlled maximum allowed rotary torque and speed, and open-hole logging unachievable using conventional wire-line tools are some of the most common limitations of this technology. However, sacrificial short-length BHAs are used to add weight to the bit and better performance in buckling load environments (Gordon et al., 2005); high-strength materials and high-torque connections are available to mitigate the negative effect of controlling certain drilling parameters; and logging-while-drilling tools are connected to retrievable BHAs (Warren et al., 2000) to obtain relevant formation data. This technology can be upgraded and optimized using either existing or modified drill stem components and following standard procedures.

In the early stages of CwD technology, the casing pipe, its connection and all conventional accessories were not designed for drilling and withstanding continuous high dynamic loads and occasional high flow rates, as opposed to drilling equipment with large internal diameter joints. In 10 years of development and implementation, key elements of CwD technology have achieved physical properties and performance never before considered by well engineers. Researchers started digging into the fatigue and mechanical performance of casing materials and connections to allow equipment to resist the dynamic drilling environment governed by high torque and high axial loads (Hossain and Homadhi, 2005, Lu et al., 2007a,Lu et al., 2007b, Teodoriu et al., 2008 and Warren et al., 2000). Services companies have invested in enhancing the performance of casing accessories such as liner hangers, float collars, cementing stage tools, and centralizers, among others, to produce casing-string components suitable for CwD (Strickler and Solano, 2007). Operators, on the other hand, are investigating the science behind the proven benefits obtained to date, such as the plastering or smearing effect (Fontenot et al., 2003). In fact, most of the available literature highlights that the plastering or smearing effect reduces lost circulation problems while drilling but clarifies that this phenomenon has not yet been proven. Regardless of when the industry will be able to represent all of the benefits of CwD with equations and reliable theorems, it is certain that CwD is already in the minds of well engineers planning drilling projects.

Benefits of Casing-While-Drilling

Regardless of the complexity of the application, most of the proven CwD benefits are mainly related to borehole stability, wellbore integrity, a reduced number of casing/liner strings, personnel safety, and overall drilling performance in on- and offshore projects. Nowadays, well engineers want to take this technology beyond traditional CwD frontiers, if they are not already. For instance, multiple consecutive top sections in which a drillable bit drills another drillable bit, extended-reach sections, non-retrievable directional CwD tools, bottom-hole drilling optimization tools, and deepening very large internal diameter casing strings are some the applications well engineers are planning.

The conventional configuration of a drilling string consists of a drill-bit, bottom-hole-assembly (BHA), and drill pipe. For a number of reasons, the BHA or drill-bit will often be removed (after reaching the final section depth) or changed by taking the entire string out of the hole as many times as necessary. For instance, directional BHAs require components that are more susceptible to damage (electronic circuits) than those used in vertical sections, and must be replaced to resume drilling. The process of tripping pipe and handling heavy BHA elements on the rig floor is tedious and accounts for up to 35% of total drilling time; this process may be even worse when unplanned events are encountered while tripping (Tessari and Madell, 1999). In some cases, unplanned events may lead to losing the well or suspending entire drilling campaigns. Moreover, it is extremely difficult to run long, rigid casing strings through poor-quality boreholes created either by pulling a conventional drill string out of the hole or by the effect of vibrations due to large borehole/pipe diameter ratio (Santos et al. 1999). Placing the casing into premature setting points leads to a need for additional casing/liner strings, which is a very expensive way to accomplish the primary drilling plan; this is one of the risks presented by drilling with casing. In most of the available technical literature, CwD is recognized as a technology suitable to drill and simultaneously case troublesome formations, reducing the risk of non-productive time and suspended drilling operations. The plastering or smearing effect is claimed to be the consequence of having a tight clearance annulus with a long, large external diameter string in continuous radial contact with the wellbore wall. This phenomenon is said to be responsible for the enhanced well integrity, as it creates a more uniform borehole and places cement sheaths in shallow, unconsolidated formations. It is also claimed that drilling with casing across formations that regularly experience severe lost circulation during conventional drilling will mechanically push the drilled cuttings into induced micro-cracks and smear produced solids against the borehole wall, creating a uniform, impermeable filter cake. As highlighted previously, more research is needed to demonstrate the veracity of this claim.

Even though CwD is considered the future drilling method in many fields worldwide, the number of companies and organizations focused on standardizing procedures and developing casing-driven integrated systems and special bottom-hole tools and accessories is less than expected. The global CwD community is still extremely undersized, and only a few companies offer limited service, while others provide some individual accessories (Kardos, 2008). In some cases, none of these services or special accessories are desirable when conventional drilling tools can be connected to casing strings and achieve the same objectives. In other words, there is no unique method or technique well engineers must select to drill with casing and cement it into place. Depending upon the application, CwD could be as straightforward as designing a conventional vertical-hole drilling string with some new considerations, or as complicated as designing a directional bottom-hole assembly. The primary objective of this document is not to describe all of the different CwD options for each drilling problem, but to highlight the fact that the benefits of drilling with casing across troublesome formations may be put at risk if proper design considerations are consciously neglected. Moreover, a CwD project must be supported by comprehensive risk assessment and analysis, regardless of the project's complexity. Knowledgeable personnel with broad drilling engineering expertise are required for a well engineer to properly design a CwD case.

RISK ANALYSIS

Risk analysis is a tool used to make decisions in the presence of uncertainties. Murtha (2000) defines risk as a "potential gain or losses associated with each particular outcome" and uncertainty as "the range of possible outcomes." In other words, risk is part of uncertainty. In cases in which there is no uncertainty, there is no risk. There are many methods used to analyze decisions when faced with uncertainties, including the probabilistic approach, Bayesian analysis, decision trees and scenarios, utility theory, portfolio analysis, and real option analysis. The probabilistic approach is generally referred in the literature as the risk analysis approach, is used to capture risk and uncertainty in the input variables, and involves running a Monte Carlo Simulation (MCS) to calculate the value of an output as a probability distribution. A detailed description of how Monte Carlo Simulations work is outlined by Murtha (2000).

Risk analysis has been applied to capture uncertainties in oil/gas field development planning, from reserves to economics (Al-Harthy et al., 2006). It has been used to determine the probability of geological success of an oil field (Otis and Schneidermann, 1997), estimate reserves (Behrenbruch, et al., 1985), and forecast production (Spencer and Morgan, 1998) and oil prices (Al-Harthy, 2007). However, the focus of this paper is the application of risk analysis to drilling operations by comparing Casing-while-Drilling technology to the more conventional approach.

Peterson and Murtha (1995) applied risk analysis to authorization for expenditures (AFEs) for drilling operation. They used statistical tools to analyze historical data to extract the parameters needed to develop a Monte Carlo Simulation model. This work showed that the probabilistic approach can yield meaningful results when applied to AFE. Peterson and Murtha (1995) presented cases using risk analysis thorough Monte Carlo Simulation of different wells and demonstrated its usefulness in predicting cost and time estimates given uncertainties in the input variables. Both works assume simple risk factors such as problem time and time added to the total drilling time; however, these risk factors were not connected to each independent drilling activity.

Thorogood et al. (2000) indicated that detailed planning incorporating risk management process in exploration wells delivered outstanding results that exceed management expectations in terms of time and cost. This risk management process includes the probabilistic approach to the AFE for drilling operations.

Akins et al. (2005) describe the probabilistic approach to estimating drilling and completion times along with cost. The authors emphasize the need to determine the level of detail required in the modeling process and select the right distribution, and highlight the effect of the central limit theorem, as well as how to reduce its impact. Cost probability curves were generated for a horizontal well for heavy oil as a case study. However, this work did not show how a risk factor could be incorporated into the probabilistic model and how it would impact the results.

Saibi (2007) carried on in the same area as Akins et al. (2005), focusing on how the probabilistic approach could be used as a prediction tool for well cost and time. He focused on the modeling of the probabilistic approach. He outlined steps and procedures that need to be followed during modeling, such as defining the distribution and building correlations. Saibi used a practical case of a horizontal well drilled in the Hassi–Messaoud field to illustrate modeling of the probabilistic approach.

Adam et al. (2009) reviewed a database of 104 wells in the North Sea for non-productive time (NPT) and showed that mechanical parent NPT, mechanical extreme NPT, and open water waiting on weather (WOW) are all statistically distinct. This work also outlined important steps and procedures that a company could use if it wanted to use risk analysis in modeling NPT. The application of this process requires a great deal of effort and personnel dedicated to the subject.

A recent survey by Hariharan et al. (2006) has shown that 91% of drilling engineers surveyed believe that there is value in using the probabilistic approach to estimate drilling time and cost. Furthermore, 46% of respondents do not use the probabilistic approach. One of the main barriers to the application of the stochastic approach, as reported from the survey, is lack of training in risk analysis and lack of tools to do the analysis. This is not a surprise, as Capen (1976) had already shown how bad engineers are at estimating risk. From our discussion with three oil and gas companies, we found out that only few drilling engineers are using risk analysis tools. Among those who are using it mostly focus on well activities time and costs and none of them incorporate risk events in the overall well cost and time.

Benefits of the probabilistic approach and the additional value provided are:

- Drilling managers lack complete future knowledge about drilling time and cost, and this is captured in the stochastic approach, which is more realistic.
- Uncertainty is a fact of life and should be incorporated in CwD drilling models. It is well known that the oil and gas business is full of uncertainties, from exploration all the way to abandonment of the field. CwD drilling operations are no exception to this rule. Cost and time are two main examples of uncertainties in drilling operations. Incorporating uncertainty reflects the reality of drilling operations.
- Decision making under uncertainty is improved by risk analysis. A good decision is one that is based on analyzing the risks and uncertainties associated with that decision in a quantified and systematic manner.
- Understanding is gained about why cost overruns exist in drilling operations and the various factors that drive these costs.
- The factors that impact cost and time the most are revealed, and specific attention can then be paid to these.

Very little work has been done on combining risk analysis and Casing-while-Drilling. In fact, we find only two pieces of work in this area: Tessari and Warren, 2006 and Houtchens et al., 2007. Tessari and Warren (2006) presented an example in which Casing-while-Drilling reduced the number of days and trouble cost through minimizing the risk of lost circulation. However no model on risk analysis was discussed nor was any analysis using the probabilistic approach presented.

Houtchens et al. (2007) presented the benefits of Casing-while-Drilling and statistics regarding the performance of this technology compared to conventional technologies. However, no risk model was used nor were any distributions or uncertainties outlined. Furthermore, we disagree with the author's statement regarding the difficulty of applying risk models to new technologies because there is little or no experience in quantifying the probabilities and consequences of events. In fact, risk analysis is more beneficial and more essential when there is little or no information. An analogy is the widespread use of Monte Carlo Simulation or the probabilistic approach to the estimation of Original Oil in Place (OOIP). These approaches are used because we have little information about the area, net thickness, porosity and water saturation, and so we tend to use risk analysis to come up with a distribution and quantify our lack of information. The output is then used to make sound and reasonable decisions given the uncertainties that exist in the input variables. Using this analogy, we see that risk analysis is a better tool to use when little information is available about a new technology such as Casing-while-Drilling. It is necessary to integrate risk analysis and Casing-while-Drilling to demonstrate the process of incorporating risk into the modeling of CwD programs and the benefits gained from this integration.

RISK ANALYSIS: MODELING APPROACH

The modeling approach to applying risk analysis to Casing-while-Drilling is expected to quantify uncertainties in time and cost and capture the impact of risk events or factors associated with CwD. The main reason for the lack of implementation of risk analysis is the need to understand how to build a model that is able to deliver output ranges that will contain the actual results. If the actual result falls outside the range that was forecast, then the model was not good (McIntosh, 2009). The following guidelines need to be followed when building such a model:

Scope of the Model

All drilling departments develop AFEs when drilling a new well. The first and most important concept for the well planner to understand is that using AFEs is not the right approach to build a risk analysis model. Even though AFEs produce the same output as the risk analysis model, the approach is different. The benefit of the risk analysis approach is that it gives more insight into the well cost and time than can be realized with a standard AFE model. This insight is captured in the actual quantification of the risks and uncertainties in the AFE model.

Risk analysis should start with AFE sections or phases, segmenting the whole well operation. Each phase or section will have its own time and cost. All of the factors contributing to a section will sum to the total well cost and time for that specific section.

One of the key questions is the level of detail that the model should have. Williamson et al. (2006) indicated that there are three levels: well level, section level, and job level. In our approach, we see the section level as the best way, as it provides the flexibility to go to well sections or to even greater detail at the job level. If the decision to be made requires the need to model these details, then the model should be able to do this; otherwise, it should not.

We also agree with Williamson et al. (2006) that the level of detail should be limited to the basic elements, such as the rate of penetration (ROP) (m/day) or material costs ($/ft). This gives the well planner the ability to trace the cause of impact in a different variable back to a basic variable.

Costs should also follow the division of the whole well into sections. The daily operating cost should be divided into two groups: those related to the overall operation or common to each section and those specific to some sections. The material cost should be divided into two groups: material costs needed for all sections and material costs related only to specific sections. The cost of the sub-sections should sum to the total well cost. The same method that is used for cost should be followed for time as well.

Moving from Deterministic to Stochastic

The key to performing risk analysis is moving from a deterministic approach, in which single values are used, to a stochastic approach in which inputs are represented by a distribution. This captures uncertainty that exists in the input variables, and the drilling manager cannot predict the exact value of a variable. Distributions for input variables can be determined in two ways:

- If prior data exist, a Probability Distribution Function (PDF) can be selected to represent the data.

- In the absence of data, a distribution may be suggested by an expert on the subject matter.

Adam et al. (2009) have discussed the details of developing a database for drilling input variables and the need to know which distribution tends to best fit the data. Williamson et al. (2006) pointed out three lessons that need to be implemented when constructing a distribution. First, avoid using the minimum and maximum of the data set as the minimum and maximum of the distribution because a minimum or a maximum beyond those of the data set could exist. Second, it should be ensured that the mean of the distribution is the same as the mean of the data set. Third, the data should be checked for outliers before the distribution is fit, or the distribution will give the wrong output. Williamson et al. (2006) have also made the point that the PDF is an approximation, and the results are unlikely to be very sensitive to small changes in the selected PDFs or parameters.

Once the input variable distributions and the equations for cost and time are defined, all of the sub-section costs and times are added up to find the total well cost and time. Monte Carlo Simulation is one of the tools used to carry out risk analysis, and is the tool used for this study. Monte Carlo Simulation involves running iterations over the sample distribution and generating a time and cost estimate, then performing iteration with different sample values; this process is repeated for the number of iterations set by the modeler. The output distribution will be smoother and have less variance for a larger number of iterations, but more iterations take more time.

Risk Factors or Events

Managers are usually qualified to identify risk factors or events that might affect drilling performance and quantify their impact with a probability of occurrence. Furthermore, they can identify ways to mitigate these risks. However, moving from deterministic to stochastic risk factors requires time and cost to be modeled stochastically, with appropriate distributions based on existing data or expert opinions.

Dependency between the time and cost for risk factors can be modeled using different methods and techniques. Al-Harthy et al. (2007) discussed four methods to model correlation: Iman–Conover, regression fitting, Envelope, and the Copula method. The Iman–

Conover is the method used in software such as Risk™ and Crystal Ball™. Regression fitting entails finding the best-fit curve using simple regression analysis. The envelope method draws lines, or an envelope, around the structure of the correlation and tries to mimic the same correlation during simulation. Finally, the copula method uses different types of copula that are able to capture different correlation structures. The right method is the one that is able to reflect the structure of the correlations in the existing data. We understand that data to support correlation may not exist, in which case any of the above methods could be used. The magnitude of correlation could be based on the opinion of an expert in the subject matter.

In addition to model dependencies, a link between the CwD risk factors and the impacted sections need to be established. One risk factor might impact one or more sections in the drilling activities. The established link should be dynamic; as the risk factors for time and cost change, all of the affected sections should change accordingly.

Model Output

The model output is as important as the model input. Traditional AFE focuses only on well time and cost. The risk analysis model developed has six outputs:

- Well cost: this metric focuses on the cost not related to risk factors.
- Well time: this metric focuses on the time not related to risk factors.
- Risk cost: only accounts for the cost of risk factors.
- Risk time: only accounts for the time of risk factors.
- Total well cost: includes well cost and risk cost.
- Total well time: includes well time and risk time.

This division of metrics allows managers to see the impact of well time and cost without risk factors and the added impact of risk factors to the total well time and cost. Using risk cost and risk time as separate metrics provides information about which risk factors have more significant impacts than others; as mentioned before, each risk factor may impact more than one section. Designing these metrics in this way provides more insight into what factors contribute heavily to the total time and cost of the well, ensuring better understanding and better decision-making.

CONCLUSIONS

- CwD techniques have been implemented in thousands of vertical and deviated drilling sections to mitigate wellbore instability problems and enhance drilling performance, contributing to reduced well delivery time and cost along with encouraging HSE statistics.
- Well engineering teams are taking CwD beyond its current frontiers. Drilling engineers are designing CwD programs in which key components will reach operational conditions never before expected.
- Regardless of the complexity of CwD programs, the benefits of drilling with casing may be at risk if suitable design considerations are neglected.
- Incorporating stochastic risk analysis with appropriate distributions in CwD programs, and any other drilling projects, will allow drilling managers to identify the risk factors that most heavily impact the well delivery time and cost, providing the necessary platform to improve decision-making in the presence of uncertainties.
- The integration of the CwD technology with risk analysis, and how this integration is pragmatically accomplished, was described in this document.

ACKNOWLEDGEMENT

We would like to thank John McIntosh, managing director of Eikos project systems and Dr. Schubert from Texas A&M University, and Prof. Sonia Avila from Romulo Gallegos University for reviewing the paper and their valuable comments.

REFERENCES

1.   Adam, A., Gibson, C., Smith, R., 2009. Probabilistic well time estimation revisited. SPE/ IADC 119287 presented SPE/IADC Drilling Conference and Exhibition held in Amsterdam, The Netherlands, March 17–19.

2.   Akins, W., Abell, M., Diggins, E., 2005. Enhancing drilling risk and performance management through the use probabilistic time and cost estimating. SPE/IADC 92340 presented at SPE/ IADC Drilling Conference held in Amsterdam, The Netherlands, February 23–25.

3.   Al-Harthy, M., 2007. Stochastic oil price models: comparison and impact. Eng. Econ. 52 (3), 269–284.

4.   Al-Harthy, M., Khurana, A., Begg, S., Bratvold, R., 2006. Sequential and systems approaches for evaluating investment decisions: influence of functional dependencies and interactions. J. Aust. Petrol. Explor. Assoc. 46 part 1, 7–11 May, 2006.

5.   Al-Harthy, M., Begg, S., Bratvold, R., 2007. Copulas: a new technique to model dependence in petroleum decision making. J. Petrol. Sci. Eng. 57 (2007), 195–208.

6.   Avery, M., Stephens, T., Al-Hadad, A., Turki, M., 2009. High-angle directional drilling with 95/8-in. Casing in offshore Qatar. SPE/IADC 119446 presented at SPE/IADC Drilling Conference and Exhibition in Amsterdam, The Netherlands, March 17–19.

7.   Behrenbruch, P., Turner, G., A, B., 1985. Probabilistic hydrocarbon reserves estimation: a novel Monte Carlo approach. Paper SPE 13982, Aberdeen, Scotland. September 10–13.

8.   Borland, B., Watts, R., Lesso, B., Warren, T., 2006. Designing high-angle/casingdirectionally-drilled wells. OTC 18374presented at the 2006 Offshore Technology Conference, Houston, May 1–4.

9.   Capen, E.C., 1976. The difficulty of assessing uncertainty. SPE-AIME, pp. 843–850. August,.

10. Fontenot, K., Highnote, J., Warren, T., Houtchens, B., 2003. Casing drilling activity expands in South Texas. SPE/IADC79862 presented at the SPE/IADC Drilling Conference in Amsterdam, The Netherlands. February 19–21, 2003.

11. Gordon, D., Billa, R., Weissman, M., Hou, F., 2005. Underbalanced drilling with casing evolution in the South Texas Vicksburg SPE 84173. SPE Drilling Completion J. 20 (2) June.

12. Gupta, Y., Banerjee, S., 2007. The application of expandable tubular in casing-whiledrilling.

13. SPE 106588 presented at the Production and Operations Symposium in

14. Oklahoma City, Oklahoma, U.S.A. March 31–April 3. Hariharan, P., Judge, R., Nguyen, D., 2006. The use of probabilistic analysis for estimating drilling time and costs while evaluating economic benefits of new technologies. SPE/IADC 98695 presented at the SPE/IADC Drilling Conference held in Miami, Florida, U.S.A, February 21–23.

15. Hartsema, K., Aliko, E., Campos, J., Clark, L., Delgado, F., Folling, P., Wingate, J., 2007. PDC casing drilling improves HS&E, cuts drilling costs—West Africa. SPE/IADC 105595 presented at the SPE/IADC Drilling Conference in Amsterdam, The Netherlands. February 20–22.

16. Hossain, M., Homadhi, E., 2005. Mathematical modeling for the investigation of fatigue life of casing-while-drilling deviated wells. Research Report No.32/425. King Saud University, College of Engineering, Research Center. November.

17. Houtchens, B., Foster, J., Tessari, R., 2007. Applying risk analysis to casing-while-drilling. SPE/IADC 105678 presented at the SPE/IADC Drilling Conference held in Amsterdam, The Netherlands, February 20–22.

18. Kardos, M., 2008. Drilling with casing gains industry acceptance. The American Oil & Gas Reporter. Special Report: Drilling Technology. April 2008.

19. Lu, Q., Hannahs, D., Buster, J., Coe, D., Langford, S., 2007a. Casing drilling connection qualification for shell rocky mountain project. SPE108141 presented at the Rocky Mountain Oil & Gas Technology Symposium in Denver, Colorado, U.S.A. April 16–18, 2007.

20. Lu, Q., Hannahs, D., Wu, J., Langford, S., 2007b. Connection performance evaluation for casing-drilling application. OTC 18495 presented at the Offshore Technology Conference in, Houston, Texas. April 30–May 3, 2007.

21. McIntosh, J., 2009. Reality checks for well time and cost forecasts. J. Petrol. Technol. 44–49 November issue.

22. Murtha, J., 2000. Decisions involving uncertainty, an risk tutorial for the petroleum industry. Palisade Corporation.

23. Otis, R., Schneidermann, N., 1997. A process for evaluating exploration prospects. AAPG Bull. 81 (7), 1087–1109.

24. Peterson, S., Murtha, J., 1995. Drilling performance predictions: case studies illustrating the use of risk analysis. SPE/IADC 29364 presented SPE/IADC Drilling Conference Amsterdam, The Netherlands, February 28–March 2.

25. Saibi, M., 2007. A probabilistic approach for drilling cost engineering and management, case study: Hassi–Messaoud Oil Field. SPE/IADC 107211 presented at the SPE/IADC Middle East Drilling Technology Conference held in Cairo, Egypt, October 22–24. Santos, H., Placido, C., Wolter, C., 1999. Consequences and relevance of drill string vibration on wellbore stability. SPE/IADC 52820 presented at the SPE/IADC Drilling Conference in Amsterdam, The Netherlands. March 9–11.

26. Spencer, J., Morgan, D., 1998. Application of forecasting and uncertainty methods to production. Paper SPE 49092 presented at the SPE Annual Technical Conference and Exhibition, New Orleans, Louisiana. September 27–30.

27. Strickler, R., Solano, P., 2007. Cementing considerations for casing-while-drilling operations: case history. SPE/IADC 105413 presented at the SPE/IADC Drilling Conference in Amsterdam, The Netherlands. February 20–22.

28. Teodoriu, C., Ulmanu, V., Badicioiu, M., 2008. Casing fatigue life prediction using local stress concept: theoretical and experimental results. SPE 110785 presented at the SPE Western Regional and Pacific Section AAPG Joint Meeting, Bakersfield, California, USA. March 29 - April 2, 2008.

29. Tessari, R., Madell, G., 1999. SPE/IADC52789 presented at the SPE/IADC Drilling Conference in Amsterdam, The Netherlands. March 9–11, 1999.

30. Tessari, R., Warren, T., 2006. Drilling with casing reduces cost and risk. SPE 101819 presented at the SPE Russian Oil and Gas Technical Conference and Exhibition held in Moscow, Russia, October 3–6.

31. Thorogood, J., Hove, F., Loefsgaard, D., 2000. Risk management in exploration drilling. SPE 61038 presented at the SPE International Conference on Health, Safety and the Environment in Oil and Gas Exploration and Production held in Stavanger, Norway, June 26–28.

32. Warren, T., Angman, P., Houtchens, B., 2000. Casing drilling application design considerations. IADC/SPE 59179 presented at the IADC/SPE Drilling Conference in New Orleans, Louisiana. February 23–25, 2000.

33. Williamson, H.S., Sawaryn, S.J., Morrison, J.W., 2006. Monte Carlo techniques applied to well forecasting: some pitfalls. SPE Drilling and Completion, September issue, p 216–227. SPE 89984 originally presented at the 2004 SPE Annual Technical Conference and Exhibition, Houston, September 26–29.

Inherently Safer Sustained Casing Pressure Testing for Well Integrity Evaluation

Tony Rocha-Valadez[a], Ray A. Mentzer[a], A. Rashid Hasan[b], and M. Sam Mannan[a]

[a]Mary Kay O'Connor Process Safety Center, Artie McFerrin Department of Chemical Engineering, Texas A&M University, College Station, TX 77843-3122, USA

[b]Harold Vance Department of Petroleum Engineering, Texas A&M University, College Station, TX 77843-3122, USA.

ABSTRACT

This paper presents the use of a model to predict sustained casing pressure (SCP), from early pressure buildup data, as a basis for inherently safer well integrity testing. Inherently safer principles aim to eliminate or reduce the hazards by design rather than by using protective features. SCP, a well integrity issue exhibited in many wells, is any measurable pressure that rebuilds after being bled down and attributable to causes other than artificially applied pressure or temperature fluctuations in the well. Intrusion of gas, resulting in SCP, can occur because of poor cement bond in the casing or cement deterioration. Gas entering the annulus migrates to the wellhead and represents a hazard due to increased wellhead pressure and the gas inventory. Compromised well integrity can have catastrophic consequences on both environmental and safety aspects.

Most regulations require the monitoring, testing and, eventually, the elimination of SCP. However, test data analysis is predominantly qualitative and limited to arbitrary criteria. Due to the high percentage of wells that present SCP, accurate, safe and preferably fast testing methods are needed. This paper implements an analytical model, rooted in the transport processes and thermodynamics of the system, to predict pressure profiles and gas accumulation during SCP testing from early-time pressure buildup data. The amount of gas accumulated during different testing criteria, being 1) current practices and 2) early diagnostic by the analytical model, is calculated and compared. Results show that using the analytical model as a predictive tool, testing time is reduced significantly, thereby limiting the amount of gas accumulated and reducing the risk. This makes the testing procedure inherently safer as well as less time consuming.

INTRODUCTION

The uncontrolled increase of casing pressure represents a hazard that can diminish the integrity of a well and simultaneously increase the risks associated with its operation. When the pressure increase is caused by physical intrusion of hydrocarbons the phenomena is termed Sustained Casing Pressure. SCP can occur because of mechanical or thermal stresses, or can be attributed to a poor cement job. The latter can compromise the integrity of the cement allowing gas to migrate through it into the casing and then accumulating in the wellhead (Goodwin and Crook, 1992 and Jackson and Murphey, 1993). SCP is indicative of compromised well integrity; the risks of not achieving well integrity can range from the activation of rupture disks to a blowout of oil or gas. Hence, SCP in one or more casing strings dictates wellbore safety during hydrocarbon production and may indicate the need for temporary abandonment (Rocha-Valadez, Hasan, Mannan, & Kabir,

2014). In addition to oil or gas producing wells, quantifying the CO_2 leakage rate during its sequestration has gained considerable interest recently because of possible emissions into the environment. Several studies (D'Alessio et al., 2011, Huerta et al., 2009, Loizzo et al., 2011, Tao et al., 2012 and Tao et al., 2010) examine this subject and concur that the CO_2 leakage process is fundamentally similar and can lead to SCP phenomena with similar risks.

Studies submitted to the Mineral Management Service (MMS), currently the Bureau of Ocean Energy Management, Regulation and Enforcement (BOEMRE) (Bourgoyne et al., 2000 and Wojtanowicz et al., 2001), show that the casings most heavily affected, by SCP problems in the US, are the production and intermediate casings. In the report by Wojtanowicz et al. (2001), 85% of the analyzed wells presented SCP problems in at least one casing.

Besides the studies and regulations in the United States, issues with casing integrity have been documented in other countries. For instance, a systematic well integrity study byWatson and Bachu (2009) showed that over 64% of 20,725 wells in a tested area of Alberta, Canada experienced leakage through the surface-casing vent. The major contributing cause to these leaks was cited to be poorly cemented casings.

A study of gas wells by Zhu, Lin, Zeng, Zhang, and Wang (2012), performed in Chongqing, China, showed that 92.3% of the studied wells exhibited SCP problems, most of them in two of more casings. Vignes and Aadnoy (2010) explored the well integrity issues of Norway's offshore setting involving 406 wells. Their findings suggested that 18% of the wells have had integrity failures, issues, or uncertainties and 7% of these were shut-in because of well integrity issues.

Current Federal Regulation, 30 CFR 250, dictates the mandatory monitoring, testing and corrective action for casings that present substantial SCP problems. Additionally, the MMS/BOEMRE, provided general guidance in the development of the American Petroleum Institute (API) recommended practice for Annular Casing Pressure Management for Offshore Wells (API-RP 90). The current regulations and guidelines for SCP testing, although useful, are mostly qualitative and limited to arbitrary criteria relating to pressure buildup. For example, buildup tests and monitoring are bound to stop when the Maximum Allowable Wellhead Operating Pressure (MAWOP) is reached or the pressure stops increasing (reaching a stabilized casing pressure).

Fig. 1 shows one of the flow diagrams for decision making for SCP as described by API-RP 90 (API, 2006). Waiting for those cycles to reach an end point and then make a diagnostic decision might not be the best approach. First, reaching the MAWOP can lead to accumulating large amounts of gas in the annular section of the wellhead which, if released, could have severe consequences from a safety and environmental risk perspective. Second, waiting until the pressure stops increasing, assuming the MAWOP is not reached, might imply that not much gas is present; however, the time to reach this stabilized casing pressure may take months during which time the annular components of the well could be unusable, with a consequent possible negative impact on production.

This paper presents an inherently safer approach to SCP testing using an analytical model as predictive tool that considers early-time data to predict future pressure profile, gas accumulation and determine a seepage factor which is akin to permeability. The amount of time required for testing, pressure buildup and gas accumulation are calculated and compared.

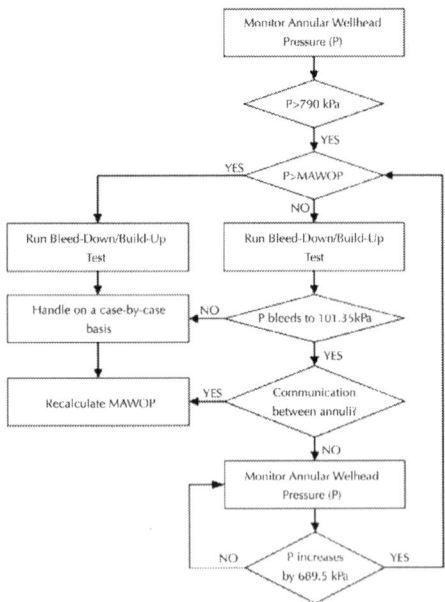

Figure 1: Flowchart for SCP decision making.

INHERENT SAFETY PRINCIPLES

Inherently safer design is an approach proposed by Kletz (1978). The concept involves a fundamental approach to hazard management that emphasizes reducing or completely avoiding the hazards at the source instead of requiring protective barriers or management systems to control them (Kletz, 1998). The inherent safety principles are: 1) Minimize (reduce the quantity of hazardous material), 2) Substitute (replace a hazardous material or process with a safer one), 3) Moderate (operate at less severe conditions or change the design and operation to minimize the effects of an incident), and 4) Simplify (avoid complexities in the system that can lead to human error or increase the probability of failure). Several authors agree that the benefits from inherently safer principles are maximized when applied as early in the process as possible (Crawley, 1995, Mannan, 2012 and Warwick, 1998) and should come first in hierarchy before prevention systems, mitigation and response (Mannan, 2012). However, the application of inherently safer principles is not limited to the design stage and should be considered and pursued at any stage of the lifecycle of the process (Khan & Amyotte, 2002).

There has been work done to implement the inherent safety principles in the oil and gas industry to reduce the risks. The main hazards on offshore installations, according to Khan and Amyotte (2002), are the process fluids and processing operations, the sea environment, and the process links between the reservoir and other installations. A clear example of inherently safer design is the automation of drilling-systems (Hansen and Abrahamsen, 2001, Kamphorst et al., 1999 and Macpherson et al., 2013) which minimize human involvement and proximity to the hazard; therefore minimizing the risk to the drilling crew. Casing while drilling is another inherently safer technology that, regardless of the complexity of the application, the proven benefits are related to borehole stability, wellbore integrity, a reduced number of casing/liner strings, personnel safety, among other advantages (Sánchez and Al-Harthy, 2011 and Sanchez et al., 2012). An example of simplification from inherently safer principles that reduces the risk between the reservoir and other installations includes the design of a structure for easy inspections which can result in costs of approximately 10% of the predicted savings in inspection costs (Hill & Bhavsar,

1996). The previous examples show that inherently safer designs and technologies, when developed and evaluated properly, are effective tools that not only reduce the risk and improve the overall process safety, but also improves the productivity or reduces operating or testing costs. It is important to mention that inherently safer principles could also, and should preferably, be applied in early design stages to prevent SCP from occurring, therefore eliminating the need to test for and quantify the problem; for example, by using self-healing cement when placing the casings. These self-healing cements can either be activated when in contact with hydrocarbons (Le Roy-Delage et al., 2010) or without any fluid contact (Reddy, Liang, & Fitzgerald, 2010). Nonetheless, a high percentage of wells around the world have already developed SCP and could benefit from this type of testing approach.

METHODOLOGY

In order to improve testing times and diagnostics for pressure buildup testing we use an analytical model developed by Rocha-Valadez et al. (2014) that has been shown to provide similar results to existing numerical methods and field data; the model is described in the Model Formulation portion of this section and has the advantage of having an analytical solution from which the time to reach a certain pressure and the gas flow rate can be estimated. In the proposed methodology we evaluate the risk from following the current practices as well as the risk from using the model described in this paper. The methodology consists of quantifying the amount of gas that is accumulated in the casinghead for three different time periods: 1) Until the pressure is at near stable conditions, 2) Until the MAWOP is reached (if it ever is reached), and 3) Until the model can predict cement seepage factor within a 95% confidence interval. The gas contained in the casinghead, due to SCP or other sources, represents a safety and/or environmental risk; particularly if the gas is flammable. Most of the gases bled-off from the annulus are either purged to atmosphere or sent to a flare. The importance of proper purging procedures and use of flares to manage the risk from gases accumulated in the annuli is critical. If an unintended release of this gas was to occur there are several possible outcomes; as described by the Center of Chemical Process Safety (CCPS, 2000) and summarized in Fig. 2. The fifth column of Fig. 2

shows the type of injury that could result from each scenario as well as the causative variables that will determine the impact of the accident. As a case study for this work we will assume that the gas contained in the annuli is released under conditions that achieve the worst-case scenario outcome, which would be an explosion. To have a sense of the consequence component of the risk from accumulated gas, the mass of the gas is converted into an equivalent mass of TNT that would release a comparable amount of energy if an explosion would occur under certain conditions; the methodology followed is presented in the Peak Overpressure Estimation section. Fig. 3 illustrates the methodology followed to compare the SCP testing methods discussed previously.

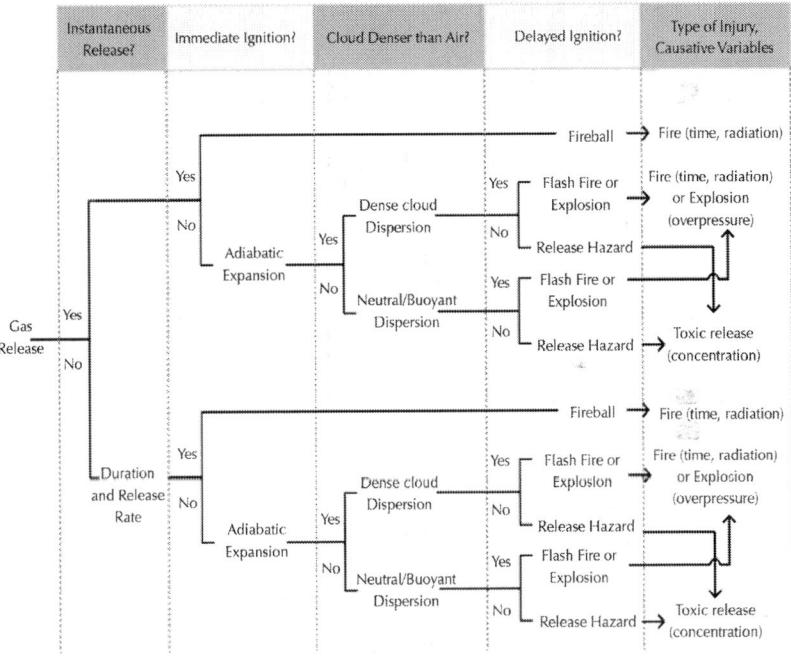

Figure 2: Event tree of gas release.

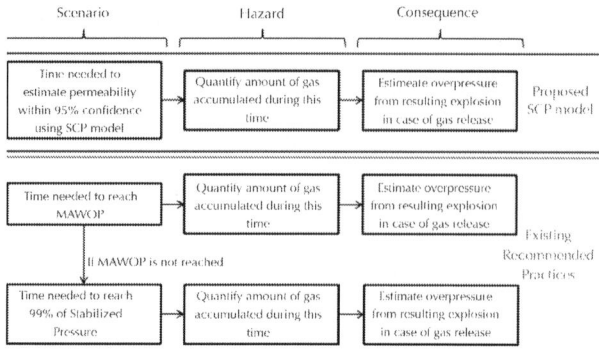

Figure 3: Hazard and consequence comparison flowchart for SCP testing.

SCP Model Formulation

The model described by Rocha-Valadez et al. (2014) is used as predictive model in this work. As sketched in Fig. 4, the system contains a gas cap of length L_g, a mud column of length L_f, and a cemented section of height L_c. The gas from the formation seeps through the cement and percolates up the mud column into the gas cap, compressing the mud column and accumulating at the top due to buoyancy. The model's implicit assumption is that the well produces at a stable rate to ensure the fluctuating tubular flow condition does not induce any heat transfer that will influence the outcome of the SCP test.

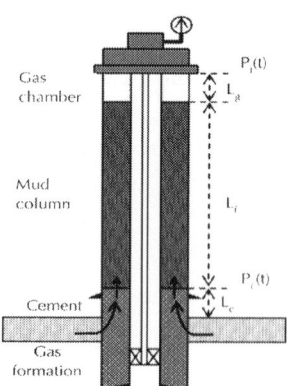

Figure 4: Cement/mud annular system schematic.

At the beginning of the test, the casing is fully open and the pressure is bled to a certain initial pressure, p_o. The instant the Annulus-A casinghead valve is closed, the gas bubbles will percolate up the mud column. The assumptions made by Rocha-Valadez et al. (2014) were that the gas leaking through the cement will move up the mud column and will displace the same amount of gas from the top of the mud column and into the gas cap implying that the gas emerging out of the cement will affect the gas cap instantaneously. The gas cap volume will increase with time given the slight compressibility of the mud and the mud-column length will decrease slowly over time. However, the pressure exerted by the mud column will remain constant since the mud is enclosed in the annulus and the mass of the mud remains constant.

The gas law, $n = pV/(ZRT)$, is used to relate gas influx rate to the rate of change of gas pressure in the gas cap. The change in the gas cap volume is accounted for by the compressibility of the mud column. Combining the material balance with the volume change of gas expansion and mud compressibility leads to the differential equation governing the transient behavior of casing pressure, which was as derived byRocha-Valadez et al. (2014) and is reproduced below:

$$\frac{dp}{dt} = \frac{\frac{0.003164 k_s A T_{wh}}{L_c \mu_g ZT}\left(p_f^2 - \left(p + 0.052\rho_m L_f\right)^2\right)}{V_i + c_m V_m p\left(1 + \frac{1}{1+c_m p}\right)} \tag{1}$$

The model developed by Rocha-Valadez et al. (2014) was derived for application in the oil and gas industry. Thus, the constant in Eq. (1) is for the case where customary oil field units are used where, k_s is in md, μ is in centipoise, pressures are in psi and volumes are in ft^3. However in this paper, input parameters and results are presented in SI units. The pertinent conversion factors, from SI to oil field units, used in the analytical model are given, in brackets, in the nomenclature section. The analytical solution of Eq. (1), as described by Rocha-Valadez et al. (2014) is expressed by the following equation in terms of the independent variable, t:

$$t = \frac{\{(\alpha - 1)V_i - \alpha\beta V_m\}\tanh^{-1}\left(\frac{p+b}{p_f}\right)}{p_f d(\alpha - 1)}$$

$$+ \frac{c_m V_m \left\{2 \ln[1 + c_m p] + (2 - \alpha)\ln\left[p_f^2 - (p + b)^2\right]\right\}}{2d(\alpha - 1)} \tag{2}$$

Where:

$$b = 0.052\rho_m L_f; \quad d = \frac{0.003164 k_s A T_{wh}}{L_c \mu_g Z P_{sc} T}; \quad \alpha$$

$$= 2bc_m - b^2 c_m^2 + c_m^2 p_f^2; \quad \beta = bc_m - 1$$

Eq. (2) is written in terms of the independent time variable rather than the customary dependent variable in order to arrive at a solution more easily. The analytical model presented by Rocha-Valadez et al. (2014)allows for rapid seepage factor estimation from a given data set. Because the pressure-time expression is nonlinear, the generalized reduced gradient (GRG) method to estimate k_s was used. The objective function for the optimization was to minimize the mean-squared error (MSE), akin to the variance of the estimator (Pestman, 2009), which is defined as:

$$\text{MSE} = \frac{1}{n}\sum_{i=1}^{n}\left(\hat{Y}_i - Y_i\right)^2 \tag{3}$$

where $\hat{Y}$ is the vector of n predictions and Y is the vector of true values. Table 1 shows comparisons for models described by Huerta et al., 2009 and Tao et al., 2010 and the presented model results for three wells. As it can be observed from Table 1, the optimized parameter, k_s, is very similar to numerical methods. Fig. 5 shows the sensitivity of k_s values estimated from the limited data showing only the maximum and minimum values of k_s estimated for and the number of points needed to obtain that value. Within Fig. 5, the average error is shown; the average error is calculated by dividing the averaged k_s values of previous estimates with the k_s value for the full set of data points. Fig. 5 also helps demonstrate that the model can serve for early-time diagnostic for n $\geq$ 3 since thereafter, prediction accuracy does not increase significantly with the number of data points used to obtain the cement seepage factor, k_s.

Table 1: Study results for Gas wells from Huerta et al. (2009) and Tao et al. (2010)

Model parameters	Case 1		Case 2		Case from Tao et al. (2010)	
	Huerta et al. (2009)	Rocha-Valadez et al. (2014)	Huerta et al. (2009)	Rocha-Valadez et al. (2014)	Tao et al. (2010)	Rocha-Valadez et al. (2014)
p_f, kPa	83,020	83,020	22,068	22,068	12,824	12,824
Initial L_g, m	0	0	3.05	3.05	0	0
c_m, kPa^{-1}	4.4E−7	4.4E−7	9.4E−7	9.4E−7	5.8E−7	5.8E−7
k, md	140	147.1	4.0	4.0	0.010	0.012

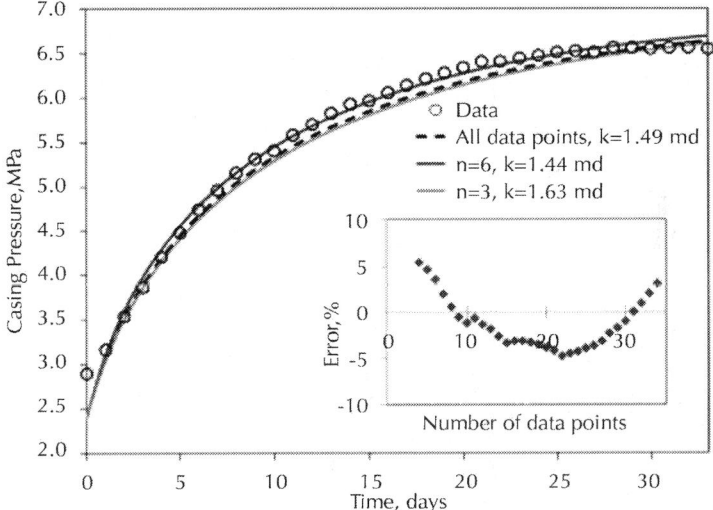

Figure 5: Pressure profiles from estimated k_s values with limited data.

Peak Overpressure Estimation

Natural gas was used as the explosive, this scenario is possible for an instantaneous release with delayed ignition or in case of a continuous release under conditions that allowed the formation of a vapor cloud that later ignited. The blast damage resulting from overpressure was estimated using an equivalent mass of TNT, m_{TNT}, and the distance from ground-zero point, denoted as r, in this case being the center top of the wellhead where the gas chamber is located. The empirically derived scaling law, as described by Crowl and Louvar (2011), is:

$$Z_e = \frac{r}{m_{TNT}^{1/3}}$$

(4)

The estimation of the peak side-on overpressure is represented by the following empirical equation:

$$P_o = P_a \frac{1616\left[1 + \left(\frac{z_e}{4.5}\right)^2\right]}{\sqrt{1 + \left(\frac{z_e}{0.048}\right)^2}\sqrt{1 + \left(\frac{z_e}{0.32}\right)^2}\sqrt{1 + \left(\frac{z_e}{1.35}\right)^2}}$$

(5)

The TNT equivalency method assumes that an explosive fuel mass behaves similar to exploding TNT on an equivalent energy basis. The equivalent mass of TNT is estimated using the following equation fromCrowl and Louvar (2011):

$$m_{TNT} = \frac{\eta m \Delta H_c}{E_{TNT}}$$

(6)

In Eq. (6), η is the empirical explosion efficiency (unitless), m is the mass of natural gas (kg), ΔH_c is the energy of explosion of the flammable gas (54,000 kJ/kg for natural gas); and E_{TNT} is the energy of explosion of TNT (4686 kJ/kg). Typical consequences from overpressure exposure are presented below inTable 2.

Table 2: Effect from overpressures (Glasstone and Dolan, 1977 and Sartori, 1983)

P_o	Effects on structures	Effect on the human body
6.9 kPa	Window glass shatters	Light injuries from fragments occur
13.8 kPa	Windows and doors blown out and severe damage to roofs	People injured by flying glass and debris
20.7 kPa	Residential structures collapse	Serious injuries are common, fatalities may occur

CASE STUDIES AND RESULTS

We will take two wells, Case Study 1 and Case Study 2; Table 3 shows relevant parameters of these wells. Using the SCP model described earlier we can estimate the seepage factor with 95% confidence as soon as 3 h after beginning testing for Case Study 1 and as soon as 4 days for Case Study 2.

Table 3: Input parameters for Case Study 1 & 2

Parameter	Case Study 1	Case Study 2
A, m^2	9.1E−3	20.7E−3
c_m, kPa^{-1}	4.4E−7	2.2E−7
k_s, md	126.3	1.496
L_c, m	1328.6	980.5
Initial L_f, m	2651.8	1960.8
Initial L_g, m	0.91	0
p_f, kPa	83,019.8	43,864.5
p_{sc}, kPa	101.35	101.35
T, K	427.8	324.4

T_{sc}, K	273.15	273.15
T_{wh}, K	299.8	288.9
Z-factor	0.86	0.92
μ, cp	0.02	0.015
ρ_m, kg/m^3	2085	1917

To use these wells as examples, and given that not all the information is available, we will make the following assumptions: the gas contained in the casinghead is natural gas, the gas contained in the wellhead from the SCP testing is released under given circumstances that result in an explosion, which represents the worst-case scenario, the peak overpressure will be estimated at a distance from the explosion, r, of 30 m and for an explosion efficiency of 5% since, according to Crowl and Louvar (2011), the most common values range between 1 and 10%. As mentioned previously, two criteria for stopping a buildup pressure test in the case of SCP are: 1) when the wellhead pressure stabilizes and, 2) when it reaches the MAWOP. Additionally, the model presented herein permits establishing a new stopping criteria, 3) when the error of the average cement seepage factor changes by less than 5%.

Pressure ratio was normalized by dividing the current pressure by the stabilized pressure; the time to reach that pressure and amount of gas in the casinghead at that given time was estimated using the analytical SCP model in this paper. Several different percentages were used ranging from the minimal value obtained from the analytical model up to 99% of stabilized pressure. Finally, the peak overpressure resulting from the ignition of that gas was estimated using the TNT equivalency method and the assumptions described above. These results are summarized in Table 4 for Case Study 1 and Table 5 for Case Study 2. Red cells are overpressures where fatalities could occur, yellow cells represent conditions at which serious injuries could be sustained and green cells are conditions where only minor injuries might occur.

Table 4: Results for Case Study 1

P/P$_{stab}$	Time hour	Gas, Kg	m_{INT}, Kg	P_o,KPa
99%	209.0	56.7	32.7	21.7

95%	118.1	52.1	30.0	21.0
90%	78.2	45.9	26.5	19.8
75%	34.5	30.4	17.5	16.6
68%	24.7	24.4	14.1	15.4
52%	9.9	17.8	7.0	11.7
49%	8.1	12.1	5.9	10.8
39%	4.0	10.2	3.1	8.5
37%	3.2	4.4	2.6	7.9

Table 5: Results for Case Study 2

P/P_{stab}	Time hour	Gas, Kg	m_{INT} Kg	P_o,KPa
99%	62.5	2.8	1.6	6.7
95%	33.5	2.5	1.5	6.5
90%	12.5	1.7	1.0	5.6
75%	9.5	1.4	0.8	5.3
73%	8.0	1.3	0.7	5.1
71%	7.0	1.2	0.7	5.0
68%	6.0	1.0	0.6	4.8
64%	5.0	0.9	0.5	4.6
60%	4.0	0.8	0.4	4.3

Case Study 1

We can observe from Table 4 that, if the stopping criterion was reaching a stabilized pressure it would take nearly 8.5 days; the consequences from an explosion after running a test for that period of time would be an overpressure of over 21 kPa which could result in death of people in buildings within 30 m. However, following the recommended API practice of stopping when reaching the MAWOP, and assuming that the MAWOP is around 20.7 MPa, we can observe that the time of testing would be reduced to a little over 24 h and reduce the peak overpressure to 15.4 kPa, indicative of significant damage and potential injuries. Finally if we run the test for 4 h, enough to successfully estimate the seepage factor, we would reduce the peak overpressure to only 8.5 kPa, where only light injuries might occur. These three criteria points can be observed in Fig. 6.

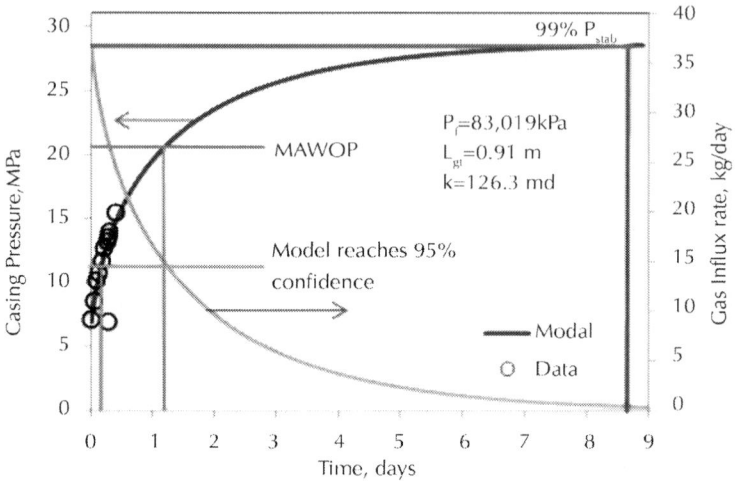

Figure 6: Wellhead pressure profile for Case Study 1.

Case Study 2

By observing Table 5, we can see that the resulting overpressures, even when reaching 99% of stabilized pressure are below 6.9 kPa, meaning that even in the worst-case scenario of release it is unlikely to have any fatalities or even major injuries. It is also unlikely that the MAWOP would be reached since the wellhead pressure is always below 7 MPa. However, as observed in Fig. 7, we can see a big difference between the time it takes to reach 99% of the stabilized pressure, 60 days, and the four days it would take to know the pressure profile by estimating the cement seepage factor with the analytical model. It would be a strong economic incentive to perform a quick test rather than waiting months to have a full pressure profile, especially if the casing should be needed for other activities which could affect production.

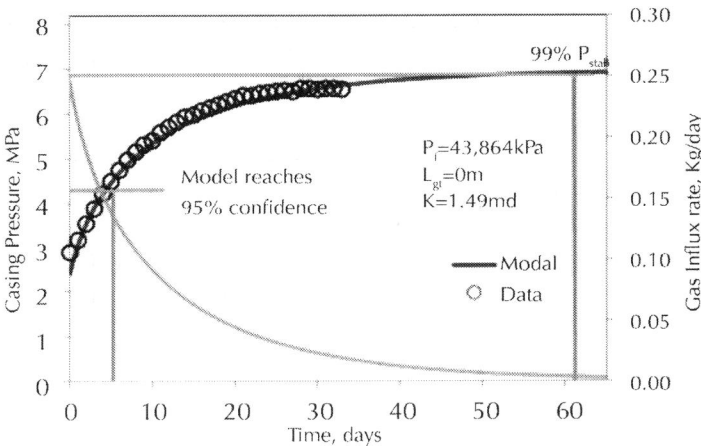

Figure 7: Wellhead pressure profile for Case Study 2.

CONCLUSIONS

The use of the SCP model, in the form of a first order, linear differential equation, allows for the estimation of the seepage factor and gas influx for wells with potential sustained casing pressure problems. Comparisons of the model against field data indicate that early-time data may be used with high accuracy for rapid estimation of the leak's severity. A comparison of the quantities of gas accumulated over different testing time periods was made. The amount of gas and the risk from the testing method proposed, showed that testing times and hazardous conditions such as high pressures and total accumulated gas can be reduced significantly. Besides practicing the inherently safer principles of minimize and moderate, this testing procedure could help reduce non-productive time. The model shows promise as a predictive tool and potential to be a basis for a standardization of SCP testing.

ACKNOWLEDGMENTS

Financial support from the Mary Kay O'Connor Process Safety Center and the Department of Chemical Engineering at Texas A&M University is gratefully acknowledged.

REFERENCES

1.   American Petroleum Institute. (August 2006). API RP 90: Annular casing pressure management for offshore wells (1st ed.).

2.   Bourgoyne, A. T. J., Jr., Scott, S. L., & Manowski, W. (2000). A review of sustained casing pressure occurring on the OCS. Final report submitted to US Department of Interior Minerals Management Service, Washington, D.C.

3.   Center for Chemical Process Safety. (2000). Guidelines for chemical process quantitative risk assessment (2nd ed.). New York: American Institute of Chemical Engineers.

4.   Crawley, F. (1995). Offshore loss prevention. Chemical Engineer, 592, 24e25.

5.   Crowl, D. A., & Louvar, J. F. (2011). Chemical process safety: Fundamentals with applications (3rd ed.). Prentice Hall.

6.   D'Alessio, P., Poloni, R., Valente, P., & Magarini, P. A. (2011). Well-integrity assessment and assurance: the operational approach for three CO_2-storage fields in Italy. SPE Productions & Operations, 26(2), 140e148. http://dx.doi.org/10.2118/ 133056-PA.

7.   Glasstone, S., & Dolan, P. J. (1977). The effects of nuclear weapons. Washington, DC.

8.   Goodwin, K. J., & Crook, R. J. (1992). Cement sheath stress failure. SPE Drilling Engineering, 7(4), 291e296. http://dx.doi.org/10.2118/20453-PA.

9.   Hansen, M., & Abrahamsen, E. (2001). Improving safety performance through tig mechanization. In SPE/IADC drilling conference. Amsterdam, The Netherlands http://dx.doi.org/10.2118/67705-MS.

10.   Hill, T., Jr., & Bhavsar, R. (1996). Development of a self-equalizing surface controlled subsurface safety valve for reliability and design simplification. In Offshore technology conference. Houston, TX http://dx.doi.org/10.4043/8199-MS.

11.   Huerta, N. J., Checkai, D., & Bryant, S. L. (2009). Utilizing sustained casing pressure analog to provide parameters to study CO_2 leakage rates along a wellbore. In SPE international conference on CO_2 capture, storage, and utilization. California, USA http://dx.doi.org/10.2118/126700-MS.

12. Jackson, P. B., & Murphey, C. E. (1993). Effect of casing pressure on gas flow through a sheath of set cement. In SPE/IADC drilling conference. Amsterdam, The Netherlands http://dx.doi.org/10.2118/25698-MS.

13. Kamphorst, G., Van Wechem, G., Boom, W., Bottger, D., & Koch, K. (1999). Casing running tool. In SPE/IADC drilling conference. Amsterdam, The Netherlands http://dx.doi.org/10.2118/52770-MS.

14. Khan, F. I., & Amyotte, P. R. (2002). Inherent safety in offshore oil and gas activities: a review of the present status and future directions. Journal of Loss Prevention in the Process Industries, 15(4), 279e289. http://dx.doi.org/10.1016/S0950-4230(02)00009-8.

15. Kletz, T. (1978). What you don't have, can't leak. Chemistry and Industry, 6, 287e292.

16. Kletz, T. (1998). Process plants: A handbook for inherently safer design (2nd ed.). Philadelphia: Taylor & Francis.

17. Le Roy-Delage, S., Comet, A., Garnier, A., Presle, J., Bulte-Loyer, H., Drecq, P., et al. (2010). Self-healing cement system e a step forward in reducing long-term environmental impact. In IADC/SPE drilling conference and exhibition. Louisiana, USA http://dx.doi.org/10.2118/128226-MS.

18. Loizzo, M., Akemmu, O. A. P., Jammes, L., Desroches, J., Lombardi, S., & Annunziatelliz, A. (2011). Quantifying the risk of CO2 leakage through wellbores. SPE Drilling & Completion, 26(3), 324e331. http://dx.doi.org/10.2118/ 139635-PA.

19. Macpherson, J., de Wardt, J., Florence, F., Chapman, C., Zamora, M., Laing, M., et al. (2013). Drilling-systems automation: current state, initiatives and potential impact. SPE Drilling & Completion, 28(4), 296e308. http://dx.doi.org/10.2118/ 166263-PA.

20. Mannan, S. (2012) (4th ed.). Lee's loss prevention in the process industries: Hazard identification, assessment, and control (4th ed.), (Vol. 1). Elsevier.

21. Pestman, W. R. (2009). Mathematical statistics (2nd ed.) (p. 232). De Gruyter.

22. Reddy, B., Liang, F., & Fitzgerald, R. (2010). Self-healing cements that heal without dependence on fluid contact: a laboratory

study. SPE Drilling & Completion, 25(3), 309e313. http://dx.doi. org/10.2118/121555-PA.

23. Rocha-Valadez, T., Hasan, A. R., Mannan, M. S., & Kabir, C. S. (2014). Assessing wellbore integrity in sustained casing pressure annulus. SPE Drilling & Completion. in press http://dx.doi. org/10.2118/169814-PA, https://www. onepetro.org/journal-paper/SPE-169814-PA.

24. Sánchez, F., & Al-Harthy, M. H. (2011). Risk analysis: casing-while-drilling (CwD) and modeling approach. Journal of Petroleum Science and Engineering, 78(1), 1e

25. 5. http://dx.doi.org/10.1016/j.petrol.2011.04.017. Sanchez, F., Said, H., Turki, M., & Cruz, M. (2012). Casing while drilling (CwD): a new approach to drilling Fiqa formation in he Sultanate of Oman e a success story.

26. SPE Drilling & Completion, 27(2), 223e232. http://dx.doi. org/10.2118/136107-PA.

27. Sartori, L. (1983). The effects of nuclear weapons. Physics Today, 36(3), 32e38.

28. Tao, Q., Bryant, S., Meckel, T. A., & Luo, Z. (2012). Wellbore leakage model for abovezone monitoring at Cranfield, MS. In Carbon management technology conference. Florida, USA http:// dx.doi.org/10.7122/151516-MS.

29. Tao, Q., Checkai, D., Huerta, N., & Bryant, S. (2010). Model to predict CO2 leakage rates along a wellbore. In SPE annual technical conference and exhibition. Florence, Italy http://dx.doi. org/10.2118/135483-MS.

30. Vignes, B., & Aadnoy, B. (2010). Well-integrity issues offshore Norway. SPE Productions & Operations, 25(2), 145e150. http:// dx.doi.org/10.2118/112535-PA.

31. Warwick, A. R. (1998). Inherently safe design of floating production, storage & offloading vessels (FPSOs). In Proceedings of offshore mechanics and arctic engineering conference. Lisbon, Portugal.

32. Watson, T. L., & Bachu, S. (2009). Evaluation of the potential for gas and CO2 leakage along wellbores. SPE Drilling & Completion, 24(1), 115e126. http://dx.doi.org/10. 2118/106817-PA.

33. Wojtanowicz, A. K., Nishikawa, S., & Xu, R. (July 2001). Diagnosis and remediation of sustained casing pressure in wells. Final report submitted to US Department of Interior Minerals Management Service, Washington, D.C.

34. Zhu, H., Lin, Y., Zeng, D., Zhang, D., & Wang, F. (2012). Calculation analysis of sustained casing pressure in gas wells. Journal of Petroleum Science, 9, 66e74. http://dx.doi.org/10.1007/s12182-012-0184-y.

Exploratory Hydrocarbon Drilling Impacts to Arctic Lake Ecosystems

Joshua R. Thienpont[1], Steven V. Kokelj[2], Jennifer B. Korosi[1, 3], Elisa S. Cheng[1], Cyndy Desjardins[3], Linda E. Kimpe[3], Jules M. Blais[3], Michael F.J. Pisaric[4, 5], and John P. Smol[1]

[1]Paleoecological Environmental Assessment and Research Lab (PEARL), Department of Biology, Queen's University, Kingston, Ontario, Canada

[2]Northwest Territories Geoscience Office, Government of the Northwest Territories, Yellowknife, Northwest Territories, Canada

[3]Department of Biology, University of Ottawa, Ottawa, Ontario, Canada,

[4]Department of Geography and Environmental Studies, Carleton University, Ottawa, Ontario, Canada, 5 Department of Geography, Brock University, St. Catharines, Ontario, Canada

ABSTRACT

Recent attention regarding the impacts of oil and gas development and exploitation has focused on the unintentional release of hydrocarbons into the environment, whilst the potential negative effects of other possible avenues of environmental contamination are less well documented. In the hydrocarbon-rich and ecologically sensitive Mackenzie Delta region (NT, Canada), saline wastes associated with hydrocarbon exploration have typically been disposed of in drilling sumps (i.e., large pits excavated into the permafrost) that were believed to be a permanent containment solution. However, failure of permafrost as a waste containment medium may cause impacts to lakes in this sensitive environment. Here, we examine the effects of degrading drilling sumps on water quality by combining paleolimnological approaches with the analysis of an extensive present-day water chemistry dataset. This dataset includes lakes believed to have been impacted by saline drilling fluids leaching from drilling sumps, lakes with no visible disturbances, and lakes impacted by significant, naturally occurring permafrost thaw in the form of retrogressive thaw slumps. We show that lakes impacted by compromised drilling sumps have significantly elevated lakewater conductivity levels compared to control sites. Chloride levels are particularly elevated in sump-impacted lakes relative to all other lakes included in the survey. Paleolimnological analyses showed that invertebrate assemblages appear to have responded to the leaching of drilling wastes by a discernible increase in a taxon known to be tolerant of elevated conductivity coincident with the timing of sump construction. This suggests construction and abandonment techniques at, or soon after, sump establishment may result in impacts to downstream aquatic ecosystems. With hydrocarbon development in the north predicted to expand in the coming decades, the use of sumps must be examined in light of the threat of accelerated permafrost thaw, and the potential for these industrial wastes to impact sensitive Arctic ecosystems.

INTRODUCTION

The Mackenzie Delta in Canada's western Arctic is underlain by significant discovered and predicted reserves of hydrocarbons`[1], but is also amongst the most rapidly warming regions globally [2]. Activities associated with the exploitation of these hydrocarbon resources, including enhanced exploration, as well as infrastructure development through the extraction, production and transmission of hydrocarbons to market, constitute an additional stressor to the freshwater ecosystems of the region. The Mackenzie Delta region is ecologically important, as identified by the establishment of the Kendall Island Migratory Bird Sanctuary in 1961, as well as being culturally significant for local indigenous communities [3]. Much recent attention has focused on oil and gas activities increasing the delivery of toxic polycyclic aromatic hydrocarbons (PAHs) into the environment [4], [5]; however the potential effects of industrial activities on aquatic ecosystems are widespread, and PAH contamination is just one example of the environmental consequences of oil and gas exploration and development.

Hydrocarbon exploration has been occurring in the Mackenzie Delta region (Fig. 1a) since the 1960s, and was particularly intense during the 1970–80s and around 2000 [6]. In the Canadian Arctic, wastes associated with the drilling of exploratory onshore hydrocarbon wells have been typically disposed of using in-ground sumps (Fig. 1b, c; [7]). These large excavations into the permafrost, usually located next to the exploratory well, are intended to act as a permanent containment location for the wastes produced during exploratory well development, including mud and rock cuttings, and drilling fluids. Drilling fluids are made up of, surfactants and detergents, as well as large quantities of highly concentrated saline solutions (primarily potassium chloride, with concentrations up to 100 g L^{-1}), which are used as a freezing point depressant during winter drilling operations [8]–[10]. Historically, a typical three-kilometre deep well, required approximately 40,000 m^3 of drilling fluid alone [7], though this volume has been reduced in more recent operations due to an improvement in recycling technologies.

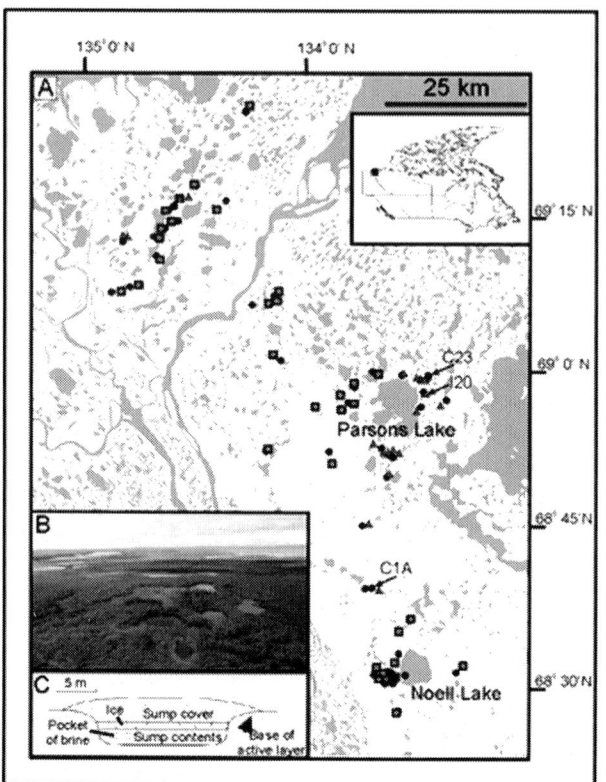

Figure 1: Map of study area, schematic of typical drilling sump and image of a degrading sump. A) The location of the 101 study lakes in the Mackenzie Delta uplands (Northwest Territories, Canada) (triangles – drilling-sump lakes; squares – thaw-slump lakes; circles – control lakes). Inset shows the region in the context of Canada. B) Image of a degrading drilling sump from the Mackenzie Delta uplands, near Parsons Lake, exhibiting significant surface and perimeter ponding. C) Generalized schematic of a drilling mud sump. A large pit is excavated into the permafrost and filled with the drilling wastes and fluids. These drilling fluids are then allowed to partially or completely freeze, and backfilled with the excavated material. The assumption is that the material will be permanently contained in the permafrost. Redrawn from [8].

More than 150 drilling sumps have been constructed in the Mackenzie Delta region since the mid-1960s [6]. As most exploration activity occurs in the winter (this practice has been required by law since 1986 in an attempt to minimize the environmental impact of drilling activities), these fluids are meant to freeze in situ, and then capped with the material excavated from the sump. It was assumed the drilling muds are permanently encapsulated within the surrounding permafrost, and that drilling sumps represent a permanent containment mechanism for materials associated with hydrocarbon exploration [7].

Pronounced climate warming in the western Canadian Arctic has serious implications for sump containment, as increasing air temperatures are increasing tall shrub cover and causing permafrost temperatures to rise [3]. Recent studies of sump integrity have observed increased conductivity beyond the extent of the sump at 74% of the sites studied [6], and as many as one third of these sites are exhibiting surface ponding, suggesting significant thaw of the sump contents has occurred [9]. Vegetation communities growing on sump caps were found to be distinct from surrounding, undisturbed areas, related to factors including drainage, active-layer depth and soil salt concentrations [10]. Increased shrub growth on sump covers enhances snow accumulation and accelerates warming of the sump and lease areas [11]. No assessment of the impacts of saline drilling wastes on nearby freshwater ecosystems exists, although the Mackenzie Delta region and Tuktoyaktuk Coastlands are amongst the most lake-rich environments in the Arctic. This is particularly important, given that many of the compounds present in drilling muds (e.g. KCl, caustic soda, barite) are known to be toxic to freshwater organisms at the concentrations common in drilling muds [7], [12], [13], and because the majority of the drilling sumps constructed are in the catchments of lakes in this water-rich landscape. Understanding the effects of drilling sump failure is essential to evaluate the relative impact of environmental stressors on freshwater in this region.

In this study, we use a combination of contemporary limnological sampling and inferences of past conditions using material preserved in lake sediments (i.e. paleolimnology) to assess the impacts of drilling fluids on the freshwater ecosystems of the Mackenzie Delta uplands region. We compare contemporary water chemistry measurements of lakes with drilling sumps in their catchments to lakes undergoing another major stressor, intense permafrost degradation in the form of retrogressive thaw slumping, as well as undisturbed control lakes with no evidence of localized disturbance. Retrogressive thaw slumps are a common form of thermokarst, which currently occur on the shoreline of approximately 10% of the lakes in the Mackenzie Delta uplands, and are increasing in size and growth rate as a result of recent warming [14].

Thaw slumping is known to result in major changes to contemporary lakewater chemistry [15], [16]and biology [17]. Our previous diatom-based paleolimnological study showed that the main biological response to permafrost thaw in the lakes of the region was related to changes in aquatic habitat availability and water clarity[18]. In this study we expand on our previous sediment-based research in order to understand the impact of hydrocarbon exploration on the lakes of this sensitive region. We used a large present-day limnological dataset of 101 sites to compare lakes with drilling sumps in their catchments to permafrost thaw-affected lakes which are known to be highly-disturbed aquatic systems in the region. Due to the fact that drilling fluids are saline, we hypothesized that if migration of drilling fluids to receiving surface waters has occurred, we should record elevated levels of major ions and conductivity in impacted lakes. Also, due to a lack of long-term biomonitoring data, a variety of sedimentary proxies were analyzed in sediment cores from three lakes in order to put the timing and nature of any limnological change in a historical context. We further hypothesized that if saline-rich wastes from drilling sumps are impacting lakes, shifts in the assemblages of sentinel biological indicators should be detected.

MATERIALS AND METHODS

One hundred and one lakes (20 drilling sump, 34 permafrost thaw slump, and 47 control lakes) in the Mackenzie Delta uplands (NT, Canada) were sampled for their present-day limnological conditions in the summers of 2005 or 2007. Thirteen measured physical and chemical variables observed above detection limits in the majority of the sites were selected and normalized. Principal components analysis (PCA) was conducted using the vegan package [19] for R [20]. Analysis of similarity (ANOSIM; [21]) and similarity percentages (SIMPER) were conducted using PRIMER v.6 in order to assess the relationships between the three a priori assigned groups, and to determine the variables that contributed to any dissimilarity.

Sediment cores were obtained from lakes I20 (68.9742°N, 133.5253°W; sump-impacted), C23 (68.9978°N, 133.5129°W; control), and C1A (68.6589°N, 133.7602°W; control) (all names unofficial) in August 2009 or July 2010 and sectioned at 0.5 cm resolution using our standard, high-resolution paleolimnological techniques [18]. Lake I20 is located downslope of a drilling sump that shows cover subsidence indicative of permafrost degradation. Lakes C23 and C1A are located nearby in similar terrain to the impacted lake I20, but with no drilling sumps in their respective catchments, and are thus classified as control lakes. Sediment age determination was conducted using ^{210}Pb and ^{137}Cs radiometric techniques (Fig. S1) [22], following which sedimentary subfossil indicators were isolated and analyzed using standard methods (diatoms:[23]; cladocerans: [24]). Relative percentage diagrams were generated using Tilia v.1.7.16 [25]. Constrained incremental sums of squares (CONISS) cluster analyses were conducted in order to identify biostratigraphic zones of change [26], with the broken stick model used to determine the number of significant zones [27]. Overall lake primary production was estimated by inferring sedimentary chlorophyll a concentrations via visual reflectance spectroscopy [28]. PAHs were extracted from wet sediments using accelerated solvent extraction (ASE, Dionex). Approval for all field-based research and the issuing of relevant scientific research licenses was conducted through the Aurora Research Institute (Inuvik, NT). No protected areas or protected species were sampled as part of this research. A more detailed description of the methods used in this study is presented in Text S1.

RESULTS AND DISCUSSION

The three a priori-defined groups (drilling sump, permafrost thaw slump and control lakes) were found to be significantly different based on their present-day water chemistry (ANOSIM, Global R = 0.307, p = 0.001), and lakes with drilling sumps in their catchments had significantly higher concentrations of chloride (Cl$^-$) than either the control lakes or lakes impacted by permafrost thaw slumps (ANOVA, df = 2, 98, F = 7.91, p<0.001; Tukey's HSD post-hoc test) (Fig. 2). The glaciogenic sediments of the Mackenzie Delta uplands are primarily derived from shale bedrock, and as such are naturally high in sulphate, but relatively low in chloride [29], [30]. As a result of permafrost thaw-slump development, the influx of terrestrially-derived materials to lakes results in a significant difference in SO$_4^{-2}$ concentration, but no significant difference in Cl$^-$ concentrations (Fig. S2, Table S1). Significantly greater Cl$^-$ concentrations in the drilling-sump impacted lakes (Fig. 2), therefore, suggests a source other than the underlying geological material, such as the wastes from drilling activity, which in most cases are brine-based (e.g. KCl) [7]. Based on similarity percentages (SIMPER), Cl^- concentrations contributed 19% of the difference between the drilling-sump and control lakes (more than any other variable), and 13% of the difference between the drilling-sump and thaw-slump lakes (Table S2). Complete comparisons of the measured physical and chemical variables are presented in Text S1. These results indicate that not only is chloride elevated in drilling-sump impacted lakes, but it is also the single most important environmental variable distinguishing sump-impacted lakes from the undisturbed control sites in the region.

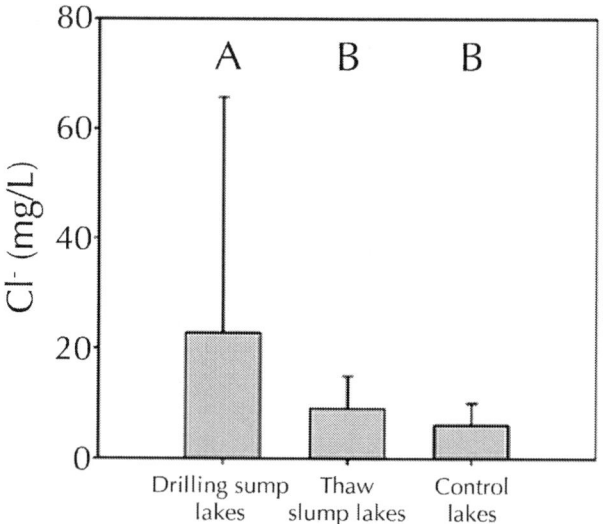

Figure 2: Mean chloride concentration in the 101 lake dataset separated into threea priori defined groups.Vertical error bars representing the standard deviation are included. Letters A and B indicate drilling-sump impacted lakes have significantly higher chloride levels than either thaw-slump affected lakes or control lakes, which are not statistically different (Tukey HSD post-hoc test, following ANOVA run on normalized environmental data).

Leaching of drilling fluids is clearly the most plausible explanation for the trends in the present-day chemical conditions observed in the drilling-sump impacted lakes in this dataset. Lakes impacted by drilling sumps also exhibited significantly higher concentrations of Ca^{2+}, Na^+, and specific conductivity compared to the control sites, though significantly less than lakes with thaw slumps (Fig. S1). Drilling fluids are the most likely source of these ions to impacted lakes, as materials such as caustic soda (NaOH) are common components [12]. In addition, as drilling fluids migrate through the active layer downslope to lake ecosystems, other ions present in soils would be translocated to the lake. This impact is exacerbated by the fact that ion-rich permafrost on top of drilling sumps and in adjacent lease areas is thawing, as active layer depth is increasing due to enhanced shrub growth [10], liberating base cations, and contributing to elevated ionic concentrations in the nearby water bodies.

Principal components analysis (PCA) was used to characterize the water chemistry variation in the 101 study lakes, taking into account the combined importance of all measured environmental variables. The first PCA axis primarily represents a response to ionic strength (Fig. 3), which is not surprising given the importance of these variables in accounting for the differences among groups. Lakes impacted by thaw slumps separate in the PCA plot from control sites primarily based on these variables [16], while the drilling-sump lakes are more widely distributed within the ordination space (Fig. 3). Some drilling-sump lakes, such as I15, I32B, I23A and I17, exhibit environmental conditions similar to sites impacted by large, active permafrost thaw slumps (Fig. 3). Drilling-sump impacted Lake I15 is chemically most similar to thaw-slump affected Lake 10B, which has an active thaw slump impacting over 25% of the lake's catchment [15], despite the lack of any natural geomorphic disturbance in Lake I15's catchment. Other drilling-sump impacted lakes, such as I1 or I12B, are chemically more similar to the control lakes, and thus the sumps in their catchments are likely providing better containment of drilling fluids (Fig. 3). Concentrations of major ions and conductivity appear to be a useful tracer for identifying the influx of materials from deteriorating drilling sumps into nearby lake ecosystems. The variability in the response of drilling-sump impacted lakes is expected, given the range of sump conditions currently observed in the region, with some sumps experiencing significant deterioration while other sites exhibit reasonable to good cover integrity [9]. Exploratory hydrocarbon drilling appears to have resulted in impacts to the freshwater ecosystems of the Mackenzie Delta uplands, in some cases more severe than large-scale, natural, localized geomorphic disturbances, such as retrogressive thaw slump development, which represent a major stressor to these ecosystems.

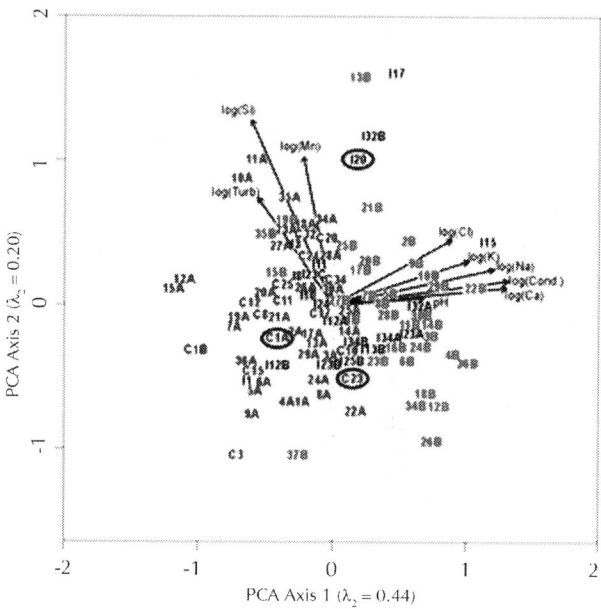

Figure 3: Principal components analysis (PCA) ordination of select chemical variables in the 101 lake dataset.Black labels represent the position of the drilling-sump lakes in the ordination space (n = 20). Red labels represent lakes impacted by retrogressive thaw slumping (n = 34). Blue labels represent control lake ordination results (n = 47). Variables were all normalized using log transformation. Arrows represent importance of given chemical variables in structuring the distribution of sites. Ordination results of study lakes I20, C23, and C1A, from which sediment cores were analyzed, are circled. Si = reactive silica.

Due to the lack of long-term biological monitoring data, paleolimnological approaches were used to assess if sump leakage has resulted in any biological changes in sentinel indicator species [31]. Multiple sedimentary proxies were analyzed in radiometrically dated sediment cores from three lakes in the Mackenzie Delta uplands: two sites near Parsons Lake, a location that has undergone significant exploratory hydrocarbon drilling over the last 50 years [6]; and one lake to the south (Fig. 1). Lake I20, impacted by a drilling sump in its catchment, shows elevated ionic water concentrations compared to the majority of the control lakes (Fig. 3), suggesting that drilling fluids may be impacting this lake. PCA (Fig. 3) revealed that some control sites, such as Lake C23, exhibited ion concentrations similar to those of some drilling sump-impacted lakes. Because these lakes have no known history of localized disturbance, we hypothesized that the elevated conductivity levels in C23 are natural, and sedimentary indicators in lake sediments should exhibit no change related to sump development. Lake C1A has conductivity levels more typical of other control lakes (Fig. 3). The analysis of I20 and two control lakes that differ in present-day conductivity will allow any changes in the drilling sump-impacted site to be placed in the context of natural chemical variability in the region.

Concentrations of polycyclic aromatic hydrocarbons did not increase following the construction of the drilling sump near Lake I20 (Text S1, Fig. S3). This suggests that, unlike the impact of large-scale extraction operations such as the Alberta oilsands [4], [5], impacts from exploratory drilling activities in this region of the Arctic do not appear related to contamination by hydrocarbons themselves. This is not surprising, given that most drilling in the delta region utilizes brine-based drilling muds (compared to oil-based), and thus sump contents proportionately contain very little hydrocarbons. Instead, any environmental impacts of these activities are likely to be related to chemicals present in the drilling fluids, notably salts.

Many species of Cladocera (Crustacea, Branchiopda) are considered poor osmoregulators, and thus we hypothesized that changes in ionic strength following drilling-sump failure would result in a shift in the species assemblage. In addition there is evidence that potassium chloride in the concentrations found in drilling fluids produces both lethal and sub-lethal effects on Cladocera in laboratory experiments [13]. Cladocera are a key component of aquatic foodwebs, and thus understanding their response to this environmental stressor is essential.

In Lake I20, an abrupt increase in the relative abundance of Alona circumfimbriata subfossils, a taxon which is known to be relatively saline tolerant [32], [33], occurred near the time of sump development in ~1972 (Fig. 4). In a survey of Cladocera in sub-Arctic lakes spanning treeline in the NWT, this taxon was most common in the highest conductivity sites [34]. While the overall magnitude of this assemblage change is relatively subtle, the timing and rapid nature of this change occurring at the time of (or very soon after) the construction of the sump suggests sump disturbance for at least this site may be primarily related to construction and abandonment practices. No comparable changes in the cladoceran assemblage of either control lake were observed coincident with, or subsequent to, the construction of the sumps near to (but not in the catchment of) those sites, despite the natural variability in ion-related water chemistry (Fig. 3). Instead, the primary changes that do occur in the control lakes are gradual and more indicative of regional warming, which has been extensive in this region and has been inferred to be impacting the cladoceran assemblages in another lake in the Mackenzie Delta [35]. The increase in A. circumfimbriata in I20 is unique when analyzed in the context of the only other regional dataset available for Cladocera in the Canadian sub-Arctic [34] (Fig. 5). Sedimentary diatom assemblages recorded no changes inferred to be as a result of drilling-sump containment loss, and instead are responding to climate warming which has been significant in this region (Text S1, Fig. S4). The lack of diatom response to these still relatively subtle changes in lakewater chemistry is not surprising. In lakes impacted by large permafrost thaw slumps, diatom assemblage shifts were inferred to be in response to modifications in aquatic habitat, and not chemical changes [18].

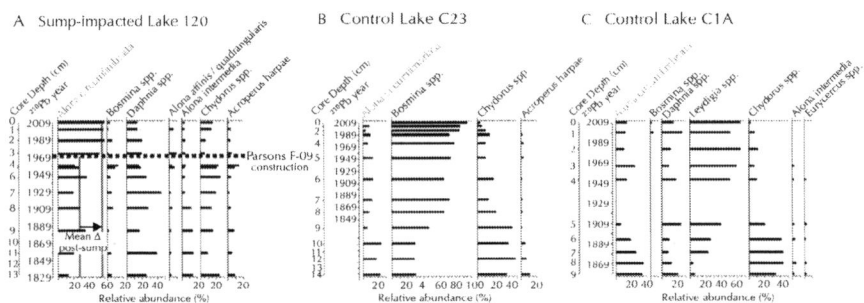

Figure 4: Stratigraphic profile of the most common cladoceran taxa from the three study lakes.Relative abundance diagrams from: lakes A) I20, impacted by drilling sump degradation; and control lakes B) C23; and C) C1A. Species assemblages (x axes) are scaled by relative abundance. Down-core sedimentary profiles (y axes) are scaled by date, based on ^{210}Pb radiometric dating techniques, with the depth in the sediment core included as a secondary axis. For all three lakes, two biostratigraphic zones were identified (constrained incremental sum of squares cluster analysis with the broken stick model)

and are plotted with the background colour of one zone in grey the other white. The known timing of construction of the compromised drilling sump near Lake I20 (industry ID: Parsons F-09) is included as a horizontal line. The vertical red lines represent the pre- and post-sump construction mean of the relative abundance of the cladoceranAlona circumfimbriata.

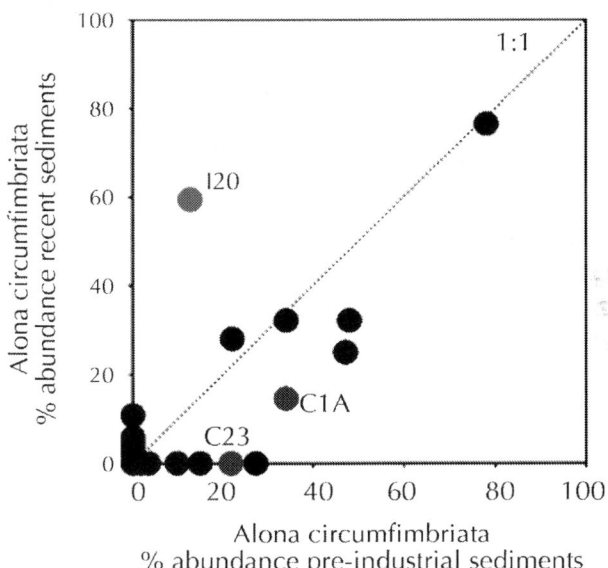

Figure 5: Change in the relative abundance of Alona circumfimbriata between present-day and pre-industrial sediment samples.Based on a comparison of the top of the sediment core (representing present-day conditions) with a bottom sample (pre-industrial) from 50 lakes in the Canadian sub-Arctic. Black circles represent 47 lakes from the only regional cladoceran survey in the Canadian sub-Arctic, spanning present-day treeline in the Northwest Territories (data recalculated from [34]). Blue circles represent the control lakes (C23, C1A) from this study. The red circle represents drilling sump-impacted Lake I20. The 1:1 line is also shown.

As the sub-Arctic and Arctic continue to warm rapidly, and permafrost temperatures increase, the likelihood of waste containment issues in drilling mud sumps will increase, potentially increasing impacts on aquatic ecosystems. Possible climate-induced increases in evaporation, as have been observed in other high-latitude regions [36], [37], could also lead to reductions in lakewater level, and further concentration of salts. This will be exacerbated by increased ionic flux due to warming and the thaw of containing permafrost. It is therefore possible that the impacts inferred from the comparison of present-day chemistry and long-term sediment records of these lakes in the Mackenzie Delta represent as yet a small component of the potential impact of drilling sumps on lake ecosystems. This is significant due to the fact that, to date, at some sites, the impacts on contemporary limnology have been as large as those associated with spectacular and conspicuous permafrost degradation. Drilling sumps represent only one example of the multiple stressors impacting Arctic lakes, which when combined with other natural and anthropogenic stressors may result in cumulative impacts to aquatic systems. If the contents of drilling sumps leach into the aquatic ecosystem due to loss of containment following permafrost thaw, the quantity and type of materials leaching into nearby lakes may increase and could result in severe, deleterious impacts on aquatic life. As the scale of hydrocarbon exploration and development in the Arctic increases in the coming decades, the potential for widespread contamination of freshwater lakes could result in cascading effects on both aquatic and terrestrial ecosystems, as well as the local indigenous communities which rely on them. In conjunction with work on permafrost conditions at drilling mud sumps [11] our data demonstrate that permafrost has not performed well as a waste containment medium, and thus if total containment of drilling wastes is a primary disposal objective, then alternative methods for waste disposal must be explored.

ACKNOWLEDGMENTS

The collection and compilation of water chemistry data from the Mackenzie Delta region was facilitated by the Cumulative Impact Monitoring Program, AANDC.

AUTHOR CONTRIBUTIONS

Conceived and designed the experiments: JRT SVK JBK JMB MFJP JPS. Performed the experiments: JRT SVK JBK ESC CD LEK. Analyzed the data: JRT SVK JBK ESC CD LEK JMB. Wrote the paper: JRT SVK JBK ESC CD JMB MFJP JPS.

REFERENCES

1. Dixon J, Morrell GR, Dietrich JR (1994) Part 1: Basin analysis. In: Dixon J, editor. Petroleum Resources of the Mackenzie Delta and Beaufort Sea. Bulletin 474Ottawa: Geological Survey of Canada. pp. 1–37.

2. ACIA (2005) Arctic Climate Impact Assessment. New York: Cambridge University Press.1042 p.

3. Burn CR, Kokelj SV (2009) The environment and permafrost of the Mackenzie Delta area. Permafrost Periglac Process 20: 83–105. doi: 10.1002/ppp.655

4. Kelly EN, Short JW, Schindler DW, Hodson PV, Ma M, et al. (2009) Oil sands development contributes polycyclic aromatic compounds to the Athabasca River and its tributaries. Proc Natl Acad Sci USA 106: 22346–22351. doi: 10.1073/pnas.0912050106

5. Kurek J, Kirk JL, Muir DCG, Wang X, Evans MS, et al. (2013) Legacy of a half century of Athabasca oil sands development recorded by lake ecosystems. Proc Natl Acad Sci USA 110: 1761–1766. doi: 10.1073/pnas.1217675110

6. Kanigan J, Kokelj SV (2010) Review of current research on drilling-mud sumps in permafrost terrain, Mackenzie Delta region, NWT, Canada. In: GEO2010: 63rdCanadian Geotechnical Conference & 6th Canadian Permafrost Conference. Richmond: Canadian Geotechnical Society. pp. 1473–1479.

7. French H (1980) Terrain, land use and waste drilling fluid disposal problems, Arctic Canada. Arctic 33: 794–806. doi: 10.14430/arctic2596

8. Dyke LD (2001) Contaminant migration through the permafrost active layer, Mackenzie Delta area, Northwest Territories, Canada. Polar Rec 37: 215–228. doi: 10.1017/s0032247400027248

9. Jenkins R, Kanigan J, Kokelj SV (2008) Factors contributing to the long-term integrity of drilling-mud sump caps in permafrost terrain, Mackenzie Delta Region, Northwest Territories, Canada. In: Kane DL, Hinkel KM, editors. Proceedings of the Ninth International Conference on Permafrost. Fairbanks: University of Alaska Fairbanks Press. pp 833–843.

10. Johnstone JF, Kokelj SV (2008) Environmental conditions and vegetation recovery at abandoned-drilling mud sumps in the Mackenzie Delta region, NWT, Canada. Arctic 61: 199–211. doi: 10.14430/arctic35

11. Kokelj SV, Riseborough D, Coutts R, Kanigan JCN (2010) Permafrost and terrain conditions at northern drilling-mud sumps: Impacts of vegetation and climate change and the management implications. Cold Reg Sci Technol 64: 46–56. doi: 10.1016/j.coldregions.2010.04.009

12. Falk MR, Lawrence MJ (1973) Acute toxicity of petrochemical drilling fluid components and wastes to fish. Ottawa: Fisheries and Marine Services, Environment Canada.108 p.

13. Utz L, Bohrer M (2001) Acute and chronic toxicity of potassium chloride (KCl) and potassium acetate (KC$_2$H$_3$O$_2$) to Daphnia similis and Ceriodaphnia dubia (Crustacea; Cladocera). Bull Environ Contam Toxicol 66: 379–385. doi: 10.1007/s00128-001-0016-z

14. Lantz TC, Kokelj SV (2008) Increasing rates of retrogressive thaw slump activity in the Mackenzie Delta region, N.W.T., Canada. Geophys Res Lett 35: L06502 doi:10.1029/2007GL032433.

15. Kokelj SV, Jenkins RE, Milburn D, Burn CR, Snow N (2005) The influence of thermokarst disturbance on the water quality of small upland lakes, Mackenzie Delta region, Northwest Territories, Canada. Permafrost Periglac Process 16: 343–353. doi: 10.1002/ppp.536

16. Kokelj SV, Zajdlik B, Thompson MS (2009) The impacts of thawing permafrost on the chemistry of lakes across the subarctic boreal tundra transition, Mackenzie Delta region, Canada. Permafrost Periglac Process 20: 185–200. doi: 10.1002/ppp.641

17. Mesquita PS, Wrona FJ, Prowse TD (2010) Effects of retrogressive permafrost thaw slumping on sediment chemistry and submerged macrophytes in Arctic tundra lakes. Freshw Biol 55: 2347–2358. doi: 10.1111/j.1365-2427.2010.02450.x

18. Thienpont JR, Rühland KM, Pisaric MFJ, Kokelj SV, Kimpe LE, et al. (2013) Biological responses to permafrost thaw slumping in Canadian Arctic lakes. Freshw Biol 58: 337–353. doi: 10.1111/fwb.12061

19. Oksanen J, Blanchet FG, Kindt R, Legendre P, Minchin PR, et al.. (2010) Vegan: Community Ecology Package. R package version 1. 17–4.

20. R Development Core Team (2010) R: A language and environment for statistical computing. Vienna: R foundation for Statistical Computing.

21. Clarke K (1993) Nonparametric multivariate analyses of changes in community structure. Aust J Ecol 18: 117–143. doi: 10.1111/ j.1442-9993.1993.tb00438.x

22. Appleby PG (2001) Chronostratigraphic techniques in recent sediments. In: Last WM, Smol JP, editors. Tracking Environmental Changes Using Lake Sediments. Volume 1: Basin Analysis, Coring, and Chronological Techniques. Dordrecht: Kluwer Academic Publishers. pp. 171–203.

23. Battarbee RW, Jones VJ, Flower RJ, Cameron NG, Bennion H, et al.. (2001) Diatoms. In: Smol JP, Last WM, editors. Tracking Environmental Changes Using Lake Sediments. Volume 3: Terrestrial, Algal, and Siliceous Indicators. Dordrecht: Kluwer Academic Publishers. pp 155–202.

24. Korosi JB, Smol JP (2012) An illustrated guide to the identification of cladoceran subfossils from lake sediments in northeastern North America: part 1—the Daphniidae, Leptodoridae, Bosminidae, Polyphemidae, Holopedidae, Sididae, and Macrothricidae. J Paleolimnol 48: 571–586. doi: 10.1007/s10933-012-9632-3

25. Grimm EC (2011) Tilia v1.6 computer program. Springfield: Illinois State Museum, Research and Collections Center.

26. Grimm EC (1987) CONISS – a FORTRAN-77 program for stratigraphically constrained cluster-analysis by the method of incremental sum of squares. Comput Geosci 13: 13–35. doi: 10.1016/0098-3004(87)90022-7

27. Bennett KD (1996) Determination of the number of zones in a biostratigraphical sequence. New Phytol 132: 155–170. doi: 10.1111/j.1469-8137.1996.tb04521.x

28. Michelutti N, Blais JM, Cumming BF, Paterson AM, Rühland KM, et al. (2010) Do spectrally-inferred determinations of chlorophyll a reflect trends in lake trophic status? J Paleolimnol 43: 205–217. doi: 10.1007/s10933-009-9325-8

29. Rampton VN (1988) Quaternary geology of the Tuktoyaktuk Coastlands, Northwest Territories. Geological Survey of Canada Memoir 423.Ottawa: Energy Mines and Resources Canada. 98 p.

30. Kokelj SV, Burn CR (2005) Geochemistry of the active layers and near-surface permafrost, Mackenzie delta region, Northwest Territories, Canada. Can J Earth Sci 42: 37–48. doi: 10.1139/e04-089

31. Smol JP (2008) Pollution of lakes and rivers: A paleoenvironmental perspective, 2nd ed. Oxford: Blackwell Publishing.383 p.

32. Bos DG, Cumming BF, Smol JP (1999) Cladocera and Anostraca from the Interior Plateau of British Columbia Canada, as paleolimnological indicators of salinity and lake level. Hydrobiologia 392: 129–141.

33. Chengalath R (1982) A faunistic and ecological survey of littoral Cladocera of Canada. Can J Zool 60: 2668–2682. doi: 10.1139/z82-343

34. Sweetman JN, Rühland KM, Smol JP (2010) Environmental and spatial factors influencing the distribution of cladocerans in lakes across the central Canadian Arctic treeline region. J Limnol 69: 76–87. doi: 10.4081/jlimnol.2010.76

35. Deasley K, Korosi JB, Thienpont JR, Kokelj SV, Pisaric MFJ, et al. (2012) Investigating the response of Cladocera to a major saltwater intrusion event in an Arctic lake from the outer Mackenzie Delta (NT, Canada). J Paleolimnol 48: 287–296. doi: 10.1007/s10933-012-9577-6

36. Smol JP, Douglas MSV (2007) From controversy to consensus: making the case for recent climatic change in the Arctic using lake sediments. Front Ecol Environ 5: 466–474. doi: 10.1890/1540-9295(2007)5[466:fctcmt]2.0.co;2

37. Smol JP, Douglas MSV (2007) Crossing the final ecological threshold in high Arctic ponds. Proc Natl Acad Sci USA 104: 12395–12397. doi: 10.1073/pnas.0702777104

Citations

CHAPTER 1

Hongping Quan, Huan Li, Zhiyu Huang, Tailiang Zhang, and Shanshan Dai, "Copolymer SJ-1 as a Fluid Loss Additive for Drilling Fluid with High Content of Salt and Calcium," International Journal of Polymer Science, vol. 2014, Article ID 201301, 7 pages, 2014. doi:10.1155/2014/201301.

CHAPTER 2

Wuchter C, Banning E, Mincer TJ, Drenzek NJ and Coolen MJL (2013) Microbial diversity and methanogenic activity of Antrim Shale formation waters from recently fractured wells. Front. Microbiol. 4:367. doi: 10.3389/fmicb.2013.00367.

CHAPTER 3

R. C. Panda, C. Lajpathi Rai, V. Sivakumar and A. Baran Mandal, "Odour Removal in Leather Tannery," Advances in Chemical Engineering and Science, Vol. 2 No. 2, 2012, pp. 199-203. doi: 10.4236/aces.2012.22024.

CHAPTER 4

G. Cordeiro, S. Dantas, L. Vasconcelos and R. Brito, "Effect of Two Liquid Phases on the Separation Efficiency of Distillation Columns," Advances in Chemical Engineering and Science, Vol. 3 No. 1, 2013, pp. 1-8. doi:10.4236/aces.2013.31001.

CHAPTER 5

I. El-Naggar, E. Zakaria, I. Ali, M. Khalil and M. El-Shahat, "Removal of Cesium on Polyaniline Titanotungstate as Composite Ion Exchanger," Advances in Chemical Engineering and Science, Vol. 2 No. 1, 2012, pp. 166-179. doi:10.4236/aces.2012.21020.

CHAPTER 6

Chakri, N. , Habes, Z. , Toubal, A. and Bendjemil, B. (2014) Synthesis of Ti3SiC2-Bicarbide Based Ceramic by Electro-Thermal Explosion. Advances in Chemical Engineering and Science, 4, 242-249. doi: 10.4236/aces.2014.42027.

CHAPTER 7

Xana Fernández-Pérez, Amaya Igartua, Roman Nevshupa, Patricio Zabala, Borja Zabala, Rolf Luther, Flavia Gili and Claudio Genovesio (2013). Innovative "Green" Tribological Solutions for Clean Small Engines, Tribology - Fundamentals and Advancements, Dr. Jürgen Gegner (Ed.), ISBN: 978-953-51-1135-1, InTech, DOI: 10.5772/55836.

CHAPTER 8

Francisco Sánchez, Mansoor H. Al-Harthy, Risk analysis: Casing-while-Drilling (CwD) and modeling approach, Journal of Petroleum Science and Engineering, Volume 78, Issue 1, July 2011, Pages 1-5, ISSN 0920-4105, http://dx.doi.org/10.1016/j.petrol.2011.04.017.

CHAPTER 9

Tony Rocha-Valadez, Ray A. Mentzer, A. Rashid Hasan, M. Sam Mannan, Inherently safer sustained casing pressure testing for well integrity evaluation, Journal of Loss Prevention in the Process Industries, Volume 29, May 2014, Pages 209-215, ISSN 0950-4230, http://dx.doi.org/10.1016/j.jlp.2014.02.012.

CHAPTER 10

Thienpont JR, Kokelj SV, Korosi JB, Cheng ES, Desjardins C, et al. (2013) Exploratory Hydrocarbon Drilling Impacts to Arctic Lake Ecosystems. PLoS ONE 8(11): e78875. doi:10.1371/journal.pone.0078875.

Index